THE RELIABILITY FRAMEWORK

Root Cause Analysis in Maintenance

THE RELIABILITY FRAMEWORK – PART 4

4

Root Cause Analysis in Maintenance

Stop failures from coming back

MARTIN VAN DEN HOUT

VULCANUS PRESS
FABRICANDO FIT FABER

Colophon

Title:	Root Cause Analysis in Maintenance: Stop Failures from Coming Back
Series:	The Reliability Framework: Nine Guides to Flawless Asset Performance
Volume:	4
Author:	Martin van den Hout
Photo's:	Depositphotos.com
ISBN:	9789083643151
Edition:	First edition: April 2026

Liability and Disclaimer

The information in this book has been compiled with great care, drawing on 35 years of practical experience in maintenance and reliability engineering. The goal is to provide maintenance professionals with practical tools for improving their maintenance processes.

Neither the author Martin van den Hout, nor Vadeo BV, nor Vulcanus Press can accept liability for the consequences of actions based on this book. Every business situation is unique and requires its own assessment of risks, costs, and benefits.

When implementing the described methods, you should take into account the specific circumstances of your organization, applicable laws and regulations, safety requirements, and manufacturer guidelines. When in doubt, consult specialists and make responsible decisions based on thorough risk analysis.

This book is intended as a practical guide and does not replace customized professional advice, adequate employee training, or sound engineering judgment.

Foreword

Anyone who works daily to keep their assets reliable knows how exceptional that profession is. Behind every running pump, every production line that keeps operating, and every network that keeps functioning, stands a skilled professional with knowledge, dedication, and an unstoppable curiosity for improvement. The maintenance professional is the silent force that makes technology perform, monitors processes, and moves organizations forward — often without anyone noticing, until something goes wrong.

Yet the world of maintenance is changing faster than ever. From TPM to Lean, from Asset Management to Reliability Engineering – the terminology shifts, but the core remains the same: reliability starts with people who know what they're doing, and who dare to learn, share, and improve.

This book is an invitation to you – the professional who looks beyond repairing what is broken. It is a compass for those who view maintenance not as a cost, but as the engine of continuity, safety, and value creation. Whether you work in industry, the public sector, or at a high-tech company, the principles shared here are universal: structured thinking, craftsmanship, and working together toward reliability.

Let this book inspire you to look beyond today. See every maintenance moment as an opportunity to learn, every problem as a reason to improve, and every asset as a system that can do better — because you want to be better.

My admiration goes out to Martin van den Hout, author and knowledge custodian with decades of experience in the maintenance and Asset Management field.

His sharp insights, thorough knowledge, and passion for sharing expertise have made an invaluable contribution to this work. Thanks to his ability to make complex subject matter clear and accessible, his contributions form a valuable bridge between theory and practice.

Because, ultimately, maintenance is not just a profession: It is a mindset!

Ellen den Broeder-Ooijevaar
Manager NVDO
(Dutch National Maintenance Society)

Table of Contents

Introduction

This book is part of a series that answers the question: "What should an organization do to make its technical installations perform as well as possible?"
I asked myself this question 35 years ago, and it has kept me constantly engaged ever since. At the time, I was working in the Maintenance & Engineering department of a Japanese manufacturing company in the Netherlands. In those days, the term "Total Productive Maintenance" (TPM) was on the rise, and more and more people were beginning to realize that simply setting up a good maintenance schedule is not enough to keep a production plant running smoothly.
Within TPM, the emphasis was strongly on involving production operators and improvements within a production department. I realized fairly quickly that this was a major step forward, but still not enough.

There were also many questions within our maintenance department that TPM didn't answer. Questions like: "Which spare parts do we need, and how many?" or "What is the best way to organize the large flow of work within a maintenance department, ranging from jobs that take a minute to projects requiring thousands of man-hours?"
In the past thirty-five years, our field has matured. TPM was incorporated into Lean Manufacturing, and from (semi-) government bodies, more and more attention was paid to technical management, under the name asset management. This ultimately led to the publication of the ISO 55000 series of standards. For the reliable operation of a production plant or other facility, the term 'Reliability' emerged as a replacement for the term 'Maintenance'. After all, reliability is the goal, while maintenance is merely one of the means.

I learned that all these schools of thought strongly resembled one another and that the differences were mainly differences in emphasis, driven by corporate culture and factors such as how many operators are directly involved in production. In the chemical industry their role is different from

that in the automotive industry, and in government organizations this role is different again.
Ultimately, many organizations have combined elements from these approaches in whatever way best suited their specific situation. Yet I learned that what applies to a production plant also applies to a network, your car, or a power plant. It doesn't matter.

It always bothered me that there were only two kinds of books about this field. The first kind describes the entire field, but only describes 'what' you should organize. They are therefore very broad, but never go into any depth. The second kind focuses on one aspect, such as workflow control or setting up preventive maintenance, and does go into depth on how to apply that one aspect.
These books have a number of things in common. They are not interconnected with one another, there are very few good ones, they are usually written in foreign languages, and they are always very thick. Often they are also written from the perspective of 'the ideal world,' in which you have an infinite number of engineers or technicians at your disposal and the most fantastic computer programs.

How This Book Is Structured

This book deliberately follows the natural steps you go through in a successful structural failure reduction effort. You can think of it as a journey in three phases, where each chapter builds on the previous one.

Laying the Foundations (Chapters 1-3) starts with the why.
Chapter 1 shows you why failures are so costly and why a systematic approach is more relevant now than ever. You will learn about the hidden costs that are 4 to 15 times higher than the visible repair costs, and you will understand how common pitfalls – such as lack of focus or premature conclusions – can block your path.

Chapter 2 covers setting up an effective registration system. Without good data, you cannot systematically set priorities. You will learn how to consistently record failure reports with clear categories and codes, so that you can later recognize patterns instead of drowning in chaotic information.

In chapter 3 you get the foundation for daily failure management. The daily failure review enables an initial triage between acute problems that you resolve immediately and structural challenges that deserve deeper analysis. You learn how to pick up early warning signals and achieve quick wins in 20 minutes per day.

From information to analysis (Chapters 4–6) is where you learn to set priorities and analyze systematically.
Chapter 4 shows you how to use Pareto analysis and risk matrices to identify the most significant losses – the 20% of problems that cause 80% of your losses. You learn to focus on what really matters instead of spreading your energy across dozens of problems at the same time.

Chapter 5 covers precise problem definition and systematic information gathering. Starting from a vague complaint like "the machine isn't working," you work toward a specific, analyzable problem. The 5W+1H checklist and Is/Is-Not analysis help you drastically narrow the search area before you begin the actual root cause analysis.

In chapter 6 you learn the methodologies for Root Cause Analysis. From simple 5-Why analyses to advanced Fault Tree methodologies – you discover which approach fits which type of problem. The 5M model ensures you don't miss any possible causes, while verification principles help you protect against premature conclusions.

From analysis to solution (Chapters 7–8) ensures that you move from theory to proven solutions.
Chapter 7 describes the verification process — how you move from suspicions to proven facts. Through systematic observations, controlled experiments, and data analysis, you transform hypotheses into reliable conclusions. You learn to distinguish between correlation and causality, and between different levels of evidence.

Chapter 8 makes the transition from proven causes to structural solutions. Systematic brainstorming with the SCAMPER methodology delivers solution directions that you evaluate with weighted decision matrices. The Four A's of change management (Accept, Align, Agree, Address) ensure that technical solutions are actually implemented and sustained. A phased approach with quick wins builds confidence for larger investments.

Each chapter closes with a natural transition to the next, so that you can see how all the elements interconnect. The story of Pedro Vargas and his team at InduPack Solutions runs as a thread throughout the book and illustrates each step with real-world examples. In the appendices you will find practical templates and tools that you can use right away.

This series consists of nine books, which together provide the answer to the previously mentioned question: "What should an organization do to make its technical installations perform as well as possible?" The books cover one topic each in a systematic, step-by-step approach. Each book connects to the other books in the series. This way, you can implement a coherent system within your company to allow your assets to perform optimally.

This is the fourth book in this series. The series consists of the following books:

Part 1: Maintenance Management

A comprehensive overview of the entire methodology. A general description of the topics covered in the other eight parts, plus several additional subjects.
ISBN 9789083643120

Part 2: Planning and Scheduling in Maintenance

Organization of the workflow of the maintenance department; including work requests, gate-keeping, work preparation, scheduling, supervision, safety, administration, improvement, and archiving.
ISBN 9789083643137

Part 3: Troubleshooting Breakdowns

The resolution of an individual failure, by a maintenance technician or an operator. Notification, finding the cause, finding a solution, repair, failure reporting, continuous improvement.
ISBN 9789083643144

Part 4: Root Cause Analysis in Maintenance

Reducing the number of failures that occur annually. Registration, statistical analysis, prioritization, root cause analysis, and implementation of solutions.
ISBN 9789083643151

Part 5: Preventive Maintenance

Setting up a preventive maintenance schedule for an entire organization, taking into account the risks and criticality of the individual assets. Translating management requirements for the entire organization into technical requirements for the assets. Mapping possible failure modes, causes, and possible maintenance actions. Economic approaches to determine which maintenance is worthwhile. Methodologies such as RCM, Industrial RCM, and pragmatic approaches.
ISBN 9789083643168

Part 6: Spare Parts Management

What types of spare parts are there? Is it worthwhile to keep them in stock? How many of each type? Organization of the warehouse and key considerations for common types of components.
ISBN 9789083643175

Part 7: Precision Maintenance

Many assets fail prematurely because they have not been assembled with sufficient precision. A technical handbook covering practical topics such as bolt tightening, fits, aligning rotating equipment, assembly with attention to balance, flange assembly, and so forth.
ISBN 9789083643182

Part 8: Engineering for Reliability

80 percent of maintenance costs and technical reliability is determined during design and construction. Project management, Early Equipment Management, involvement of operators and technicians, reliability analyses during design, FAT and SAT, Commissioning, Maintainability.
ISBN 9789083643199

Part 9: Autonomous Maintenance

Which maintenance tasks can you have operators perform, why, and how do you organize that?
ISBN 9789083706306

For a current overview of all books, additional information, and ordering options, visit:

vadeo.nl/books.html

1. Foundations

1.1. Pedro's Frustration

It was Tuesday morning at 7:30 when Pedro Vargas walked into the production floor of InduPack Solutions. The sound he heard — or rather the lack of sound — already told him enough. Conveyor line three was down again.

"Not again," he muttered, quickening his pace toward the idle line. Around the conveyor belt, controlled chaos reigned. The paper had jammed again. Sandra Hutchinson the experienced operator of line three, was hunched over the packaging machine, carefully trying to loosen stuck pieces of paper. Marco Rivera, one of the technicians, stood by helplessly watching, tools in hand.

"Pedro, same old mess again," Sandra sighed without looking up. "It's the third jam this week and it's only Tuesday. I'm running back and forth all day for these stupid jams." Marco Rivera shook his head. "I keep repairing the same thing and a week later it's broken again." Jake Jensen, the supervisor, walked over with a concerned look. "Management is getting impatient. We're missing production time again."

Pedro looked around at the idle production line. He could feel the frustration of his colleagues, but also something else – the realization that this pattern had been going on for years. Reactive repairs, no time for real improvement, the same problems recurring over and over again.

"Almost sixty hours of production loss," he said out loud. "What did you say?" Sandra asked. "That's what this conveyor belt is costing us with all those jams. 3,500 minutes per month, almost sixty hours of downtime. Plus $45,000 in repair costs."

The group fell silent. Pedro had done the calculation the previous week and the numbers were shocking. 78 jams per month, each accounting for an average of 45 minutes of downtime.

"There must be a better way," he said, more to himself than to the others. "Systematic. Focused. From reactive to proactive."

1.2. The Costly Reality

Later that morning, Pedro sat in his office with spreadsheets full of failure data. He had his mentor from his previous job on the phone – someone who had often helped him with complex technical problems.
"I only see the direct costs," Pedro explained. "Parts, labor hours for repairs. But there must be more."
"Pedro, the direct costs are only the tip of the iceberg," his mentor replied. "What does the production loss cost? The rejected products? The overtime? The missed delivery dates?"
Pedro began working through the calculations in his spreadsheet:
Direct repair costs: $45,000 per year
Production loss: 3,500 minutes at $200 per hour is $140,000 per year
Rejected products: $25,000 per year
Overtime: $18,000 per year
Customer compensation for missed deliveries: $12,000 per year
"$240,000 total," he whispered into the phone. "More than five times the direct costs."

At that moment Linda Wang, the quality engineer, knocked on his door and came in. "I heard you on the phone. Is it about the conveyor belt?"
Pedro nodded. "Linda, we always talk about repair costs, but take a look at this calculation."
Linda leaned over his screen. "And we haven't even talked about the stress on the team. Sandra looks tired. Marco is getting cynical. Jake is constantly under pressure from management."
"Four to fifteen times higher hidden costs," Pedro said. "I've read about it in trade journals. Here I see it in practice."

1.3. The Turning Point

That afternoon, Pedro called his team together for a meeting. In addition to Sandra, Marco, Linda, and Jake, Tom Davis was also present, the electrical technician.
"$240,000," Pedro began, pointing to his laptop screen on the conference table. "For one conveyor belt. One of the twelve lines we have."

Sandra let out a soft whistle. "And if this is representative of our other lines..."
"Then we're talking about millions per year," Pedro finished the sentence. "And we're not unique. This happens in thousands of companies, every day."

The team fell silent. For Pedro, this was a moment of clarity that he would later describe as his "aha moment." For years, he had seen failures as inevitable – something that simply came with the territory of production. Machines fail, you repair them, and you move on. But now he saw the reality: InduPack wasn't managing production, they were managing failures.
"We think we're a manufacturing company," he said to the team, "but in reality we're a repair company that happens to produce something as well."
Marco shook his head. "That's a hard confrontation with reality."

"And our competitors?" Jake asked. "Are they struggling with the same thing?"
"The bad ones do. But the successful companies – the real top players – broke through that years ago. For them, continuous improvement isn't a nice-to-have, it's in their DNA."
Pedro stood up and walked to the whiteboard on the wall. He wrote two columns:

Reactive Operations

- Waiting for machines to break down
- Repair and Move On
- Accepting Failures as 'Normal'
- Focus on Symptoms
- The Same Problems Over and Over

Proactive Enterprise

+ Preventing Machines from Breaking Down
+ Systematic Improvement

- Seeing Failures as Waste
- Focus on Root Causes
- Solving Problems for Good

"Look," Pedro said, gesturing toward the team, "this is where the difference between winners and losers in the industry lies. Between companies that grow and companies that barely survive. Between teams that take pride in their work and teams that only put out fires."
The team studied the board in silence. Sandra was the first to speak. "So this isn't just about our conveyor belt?"

"No, this is about who we want to be. About whether we keep fighting the same failures for years to come, or whether we finally take the step toward structural improvement."
Pedro turned to his team. "You know what the real problem is? It's not that companies don't know that failures cost money. It's that they don't believe they can do anything about it. They've tried to improve things so many times without success that they've become cynical."

"But is it possible?" Tom asked.
"It can't just be optional—it's mandatory. In a world where margins are under pressure and customers are demanding more and more, you can no longer afford the waste caused by failures. The companies that don't understand that will disappear. The companies that do understand it will dominate their market."
Pedro picked up a marker and underlined the words 'PROACTIVE COMPANY' on the whiteboard.

"But people, and this is crucial – it's not about good intentions or brainstorming sessions. It's about a systematic, disciplined approach. Focus. Priorities. And above all: learning from the mistakes that other companies have already made."
He paused for a moment and looked around at the team. "Because there are companies that have done exactly what we are now trying to do – and failed. And we need to learn from that before we begin."

The Cautionary Tale of TechCorp

Pedro wanted to tell them something about his previous employer – a story he used as a warning.

"Before I came here, I worked at TechCorp Manufacturing," he began. "There we did everything you should NOT do. We tried to tackle 127 problems at the same time."

Sandra raised her eyebrows. "127? That sounds like chaos."

"It was," Pedro confirmed. "The technical manager had organized a brainstorming session with all department heads. They made a list of all failures and problems. Each department was assigned fifteen to twenty problems."

Marco leaned back in his chair. "And what happened?"

"Every time a failure occurred, the improvement work was pushed aside. 'Fighting fires' always took priority over structural improvements. After a year, 8 of the 127 problems had been addressed, and those weren't even working."

Jake frowned. "Why didn't they work?"

"No focus, no priorities, no system. And above all: no dedicated resources. You can't expect people to solve acute and structural problems at the same time."

Tom nodded slowly. "And here we're going to do it differently?"

"Here we work systematically. A few problems at a time at most, thorough analysis, proven solutions. Focus instead of chaos."

1.4. Redefining Failures

Sandra set down her coffee cup. "A failure is just when a machine breaks down, right?"

"That's what we thought at TechCorp too," Pedro replied. "But a failure is actually any situation in which an asset doesn't meet your expectations."

Marco looked interested. "What do you mean?"

"If the conveyor belt is running, but 20% slower than normal – is that a failure?" Linda thought for a moment. "Technically, it's still working."

"But it doesn't meet our expectation of normal speed. Production loss, higher cost per product. That's also a failure."

Tom nodded. "So quality problems too? Safety risks?"

"Exactly! By defining failures broadly, we get a more complete picture of improvement opportunities. We're not just looking at downtime, but at everything that causes our assets to perform suboptimally."
Jake began to understand. "That explains why we never saw all the problems. We were only looking at the big, obvious failures." Sandra nodded in agreement. "And all those little annoyances that we consider 'normal'?"
"Those can together be larger than the major failures. That's why we're going to systematically look at everything that isn't running optimally."

1.5. The Role of the Reliability Engineer

Marco straightened his cap and looked at Pedro. "You're a reliability engineer. What does that actually mean, exactly?"
"My task is to ensure that our assets function more reliably. Primarily by structurally reducing failures."
Sandra leaned forward. "And what about us? What's our role in this?"
"You are crucial! I can't improve things systematically without your practical knowledge. Sandra, you're the first to notice when something's off. Marco, you know how machines really work."

Tom frowned. "But why do we need a dedicated person for improvements? Can't we do that ourselves?"
"Because if you have to combine this task with putting out daily fires, the structural work keeps getting pushed aside. Acute failures always disrupt the planning."
Jake nodded vigorously. "Improvements can always wait a day when there's something urgent."
"Exactly! That's why I have dedicated time to focus on structural improvements. But that doesn't mean I do it alone – we do it together."
Linda smiled. "So you facilitate, we provide the expertise?"
"Well summarized. Together we make assets more reliable."

Reactive versus Preventive versus Predictive

Jake put down his notebook. "We're mostly reactive right now, aren't we?"
"That's right. 'Fix when broken'. Waiting until something breaks down and then reacting."
Marco defended his approach. "That works, doesn't it? I fix everything fast."
"You do fantastic repair work, Marco. But reactive maintenance has its drawbacks: unexpected downtime, higher costs, secondary damage, disruption of production."

Sandra looked up from her notes. "And preventive maintenance?"
"Maintenance according to a fixed schedule, regardless of condition. Like having your car checked every 20,000 kilometers."
Tom was intrigued. "And predictive?"
"Perform maintenance only when measurements indicate it is necessary. Like taking your car in for a check because a warning light is on."
Linda connected the dots. "Where does failure analysis fit in?"
"In all three! But especially in the transition from reactive to preventive and predictive. By thoroughly analyzing failures, we discover why things fail, we can reduce the number of failures, and for the remaining failures we can better predict when they will fail again."

1.6. Discovering the 20/80 Principle

Pedro stood up and walked to the whiteboard. "I'm going to show you something that will change your world. The Pareto principle: 20% of your problems cause 80% of your losses."
Sandra looked skeptical. "What do you mean?"
"Of all the failures we have, there are probably a few that are responsible for most of the loss."
Marco nodded slowly. "I can imagine that. Some failures are worse than others."
"More than you think. At TechCorp, we discovered that three out of 127 problems were responsible for 65% of all losses."
Linda looked surprised. "And the other 124?"

"Together accounting for 35%. Just imagine: if we had solved those three problems instead of spreading ourselves across 127..."
Jake understood. "Then with two percent of the effort we would have achieved 65% of the improvement." Tom slapped his hand on the table. "That explains why our current approach isn't working. We're wasting energy on small problems."
"Exactly! Focus on the most important ones. That's what we're going to do here as well."

1.7. Human Factors – The Forgotten 50%

Sandra put down her pen and looked thoughtful. "We've bought a lot of new equipment in recent years, but the problems keep coming."
"That's because 50% of all failures are caused by human factors."

Marco became defensive. "Are you saying we make mistakes?"
"Not only errors. Think about: insufficient training, unclear procedures, fatigue, communication problems, incorrect work methods."
Tom understood where Pedro was going. "So technical solutions aren't enough?"
"Technology is half the puzzle. The other half is people, procedures, training, communication."

Jake slapped himself on the forehead. "That explains why some of the 'technical' solutions didn't work." Linda added: "We addressed the machine, but not the way people worked with it."
"Exactly! That's why we're going to look at both aspects: technology and people. A good solution integrates both." Sandra nodded slowly. "So if we really want to solve this problem, we need to look at ourselves too?"
"We look at the complete system: machine, people, procedures, environment. Together."

The First Systematic Step

Pedro put down his marker and looked at the group. "Okay team, now that we understand the basics, let's get to work systematically."
Marco leaned forward. "Where do we start?"

"With one problem. Those conveyor belt jams. I've already established that they are our biggest problem. Not all 78 types at once, but focused on what has the greatest impact."

Sandra was practical. "And how do we approach that?"
"Step by step. First we make sure we register all failures properly. Then we set priorities. Then we systematically analyze the root causes."
Jake nodded. "And then solutions?"
"Solutions that work. No quick fixes that fall apart a month later. Structural solutions that eliminate problems."
Linda looked at her watch. "How long is this going to take?"
"A few months before we see the first major results. But I promise you: if we do this right, we'll have 60-85% fewer failures within a year."

Tom let out a soft whistle. "And the costs?"
"This investment pays for itself many times over."
Sandra shook her head. "Sounds too good to be true."
"That's what I thought too. But methodology works. Not through magic, but through discipline. Step by step, we move toward structural improvements."

1.8. The Foundations of an Improvement Process

Pedro walked back to his chair and sat down. "Let me explain what a systematic failure-driven improvement process entails."
He began to enumerate, counting on his fingers:

1. "Clear definition of what is considered a failure
2. Consistent recording of all relevant failure details
3. Prioritization based on impact and frequency
4. Thorough investigation of underlying causes
5. Development of structural solutions
6. Implementation of improvements
7. Verification of effectiveness
8. Standardization of successful solutions
9. Knowledge sharing throughout the entire organization"

Marco nodded slowly. "That's a lot of steps."
"By consistently applying this process, we can achieve impressive results," Pedro explained. "We can systematically improve asset reliability, reduce maintenance costs, increase productivity, improve safety, reduce environmental risks, and build a culture of continuous improvement."

Linda was intrigued. "And does the Pareto principle play a role in that?"
"Absolutely. The 80/20 principle helps us focus on what truly matters. Approximately 20% of your problems cause 80% of the impact. By concentrating on these few most important ones, you achieve the greatest improvements with the least effort."
Tom leaned back. "And those human factors you mentioned?"
"We need to include those as well. Reducing failures is more than just technology."

Why Organizations Fail

Jake tapped his pen on the table. "You've told the story of TechCorp. Are there more patterns of how organizations fail?"
"Sure," Pedro replied. "I see four main patterns that derail organizations."
He stood up and began to speak.

"Failing to set priorities. Organizations try to tackle all problems simultaneously instead of focusing on those few problems that have the greatest impact: 'the vital few.' This leads to fragmentation of resources and few concrete results."
Sandra recognized this. "That sounds like what we do now. Every problem gets a little bit of attention."

"Second pattern: no dedicated resources. The same people are responsible for both daily problems and structural improvement. When a failure occurs, all structural work is pushed aside."
Marco nodded. "The result is that organizations continuously remain in reactive mode."

"Third pattern: symptom management. Teams focus on resolving the immediate consequences of failures, without addressing the underlying causes. As a result, the same problems keep recurring."

Tom understood. "And the fourth pattern?"
"Jumping to conclusions. Teams assume too quickly what the cause of a problem is, without thoroughly verifying it. This leads to solutions that don't work or only have a temporary effect."
Linda looked thoughtful. "I recognize all these patterns."
"Most organizations make these mistakes. The good news is that you can avoid them if you recognize them."

The Journey Begins

Jake was practical. "What's the first step?"
"Tomorrow we start setting up a proper registration system. Without good data, we can't set priorities systematically."
Marco was impatient. "And then?"
"Then daily discussions to separate acute problems from structural challenges. Then we analyze which problems have the greatest impact."
Sandra looked hopeful. "And then finally tackle those conveyor belt jams?"

"Then we'll systematically search for the root causes and develop structural solutions."
Linda smiled. "For the first time, it feels like we have a plan." Tom added, "Instead of just reacting to whatever happens."
Jake looked at the clock. "And if it works?"
"Then we roll it out to other problems. From one successful improvement to a culture of continuous improvement."
Sandra stood up. "When do we start?" Pedro smiled. "We've already started. By thinking systematically instead of acting reactively, we've already taken the first step."

The Way Forward

After the meeting, Pedro stayed behind in the conference room for a little while. He thought back to his time at TechCorp and the frustration he had felt there. Here at InduPack, he had the chance to do things differently — systematically, focused, with a dedicated team.
The conveyor belt line three would become their test case. 78 jams per month, $240,000 annual loss - but also a concrete opportunity to prove that a systematic approach worked. The team had shown confidence, despite their earlier disappointments.

Tomorrow they would begin with the real work. First, setting up a robust registration system for workflow implementation of registration and feedback. That system would form the foundation for all further improvements.

Pedro stood up and took one last look at the whiteboard covered with all the notes from the conversation. Now that the team understood the fundamentals of why failures are so costly and how organizations often fall short in their approach, they could begin building a systematic improvement process.

In chapter 2 you will read how they learn to set up a robust registration system that forms the foundation for all further improvements. This system would ensure that they collected the right data to later be able to set effective priorities and carry out targeted analyses.

The journey toward structural improvement had begun. With a systematic approach, teamwork, and perseverance, they would definitively overcome the conveyor belt jams – and in doing so, set an example for the rest of the organization.

2. Setting Up a Failure Registration System

2.1. Setting Up an Effective Registration System

Pedro looked at his laptop screen at the failure data from conveyor belt line three. What he saw was frustratingly inconsistent. He read the various notifications aloud.

"Notification 1: 'jam resolved'."

Notification 2: 'Conveyor belt line three stopped at 2:32 PM due to a paper jam in section C caused by humid weather conditions and possible wear of guide roller 7, resolved by cleaning the roller and adjusting the tension, lasted 23 minutes'."

Linda Wang, the quality engineer sitting across from him, asked, "What is the problem?"

"One notification tells me nothing, the other way too much. And look here — three different people describe the same problem in three completely different ways." Linda looked over his shoulder. "Marco writes 'mechanical defect.' Sandra writes 'paper jam.' Tom writes 'electrical sensor indicated a failure.'"

"They're talking about the same jam, but I can't find any pattern in it. Without good data, I'll never be able to determine which of those 78 monthly jams should be prioritized."

"So before we can solve problems, we first need to learn how to describe them?"

"Exactly! A doctor can't make a diagnosis either if every nurse describes symptoms differently."

Now that you understand the fundamentals of why failures are so costly and how organizations often fail in their approach, you can begin building a systematic improvement process. The first crucial step is setting up a robust registration system that forms the foundation for all further improvements.

The Enterprise Asset Management System

The following morning, Pedro organized a session with the team about their Enterprise Asset Management system. Jake Jensen, the production supervisor, Sandra from production, Marco Rivera from maintenance, and Tom Davis from electrical sat around the conference table.
"Don't we have a system?" Jake asked. "Everyone reports failures in the EAM."
"You report, but not systematically. Look here." Pedro showed his laptop screen. Marco Rivera shrugged. "I don't see a problem. I write down what was broken and how I fixed it." Sandra added, "And I report when the machine stops and when it starts again." Tom nodded, "I focus on technical details - which components, which settings."

"Everyone reports from their own perspective. That's valuable, but we're missing a common structure," Pedro explained. "What do we need then?" Linda asked. "Clear fields for symptoms, causes, and solutions. Standardized categories. And above all: consistency in how we describe things."
Jake frowned. "Does that mean more paperwork?"
"At first, yes. But if we do it right, we can see patterns that are currently invisible. And patterns mean structural solutions."

You must document every failure or other source of production loss or potential hazard in your Enterprise Asset Management System. This system ensures that you collect the right data to later set effective priorities and perform targeted analyses.
In the EAM, there is an overview of all assets in your organization, with a great deal of relevant data linked to them. This includes technical specifications, the location of the asset, and information about the manufacturer or supplier. In addition, you can record and retrieve information in the EAM about the preventive maintenance that needs to be performed on the asset, as well as all work that has been completed or planned.
For failure registration, it is relevant that you can quickly find the assets in the EAM to report a failure. After the failure has been repaired, you must record the information about the failure in the system. The maintenance

technicians must write this failure report, and it is the responsibility of the department manager to ensure that they do this in a clear manner.

2.2. Required Data in Failure Reports

A week later, Pedro led a workshop on what they needed to record at minimum with every failure report.
"For every failure, we need ten essential data points." Sandra looked skeptical. "Ten? That's going to be like writing an essay every time."

"Not if we're smart about it. Look, we need this data for several purposes."

Pedro wrote on the whiteboard:

1. Asset name and number
2. Date and time of occurrence and resolution
3. Caller and Technician
4. Duration and production loss
5. Safety and Environmental Impact
6. Symptoms
7. Cause
8. Root Causes
9. Solution
10. Follow-up Actions

Marco looked at the list. "Why both symptoms and cause?"
"The symptom is what you see: 'paper jams.' The cause is why: 'guide roller too much tension.' The root cause is why that tension exists: 'alignment lost after maintenance.'" Tom nodded slowly. "So we go from what to why to why of the why?"
"Exactly! Symptoms help with quick diagnosis. Causes help with repair. Root causes help with prevention."
The failure report serves several purposes. As a maintenance technician, you can use it to indicate, for example, that additional follow-up work is still needed to fully resolve the failure. You can also review past reports to see whether the failure you are currently working on has occurred before. That is why it is important that the title of the failure describes the most important system – essentially the notification you receive from an operator. Instead of "it doesn't work" it should be "what it doesn't do." When

you as a maintenance technician look back at the history to see whether the failure you are currently working on has occurred before, you don't yet know the cause. So it should not say "PLC card replaced" but rather "machine shuts down ten seconds after startup every time."

FAILURE REPORT		Date	Department	Asset name	ID No.
Title					
Technician(s)			Operators		
Failure start	Failure end	Production downtime	Repair start	Repair end	Hours worked
Impact	Safety [ ]	Environment [ ]	Quality [ ]	Regulatory [ ]	
Symptoms					
Direct cause					
Repair performed					
Settings / parameters after correction					
5 x Why 1			Materials description	Quantity	No.
2					
3					
4					
5					
Follow-up actions required?					
Remarks / lessons learned					

Figure 2-1 Failure report

Another important purpose of failure reports is to record information needed to definitively resolve the failure and to assess whether this is worth the cost and man-hours.

You must include the name of the reporter and the repair technician so that clarification can be requested when not everything is completely clear. Once the reliability engineer starts examining the details of a failure, all kinds of questions often arise whose answers are not included in a failure report.

The duration of the failure and its other consequences are needed to determine whether this failure should be prioritized for structural improvement. The symptoms, cause, and solution are essential to developing a structural solution for the failure.

Implementing Categories and Codes

He displayed an overview with three columns on his screen.

The Three Pillars of Failure Registration

"We describe every failure with three elements," Pedro began. "What you see, why it happened, and what you did."

Column 1: Damage Codes – What Is Visible?
SC01 - Failed (machine won't start, installation not working)
SC02 - Leak (oil, water, air leaking away)
SC03 - Loosened (bolts loose, cables detached)
SC04 - Broken Off (broken parts, cracks)
SC05 - Clogged/Contaminated (filter blocked, contamination on components)
SC06 - Wears Abnormally (faster wear than expected)
SC07 - Makes noise (unusual noise, rattling, squeaking)
SC08 - Vibrates/Moves (abnormal vibrations, movement)
SC09 - Overheating (bearings, motor, components running too hot)
SC10 - Performs poorly (low capacity, slow operation)

Column 2: Cause Codes – Why did this happen?
OC01 - Normal Wear (expected service life reached)
OC02 - Operating Error (misuse, incorrect operation)
OC03 - Maintenance Backlog (Late, Missed, Not Performed)
OC04 - External cause (power outage, weather, contamination)
OC05 - Material Defect (defective part, poor quality)
OC06 - Assembly Error (incorrectly assembled, poorly adjusted)
OC07 - Overload (excessively loaded, limit exceeded)
OC08 - Unknown (cause cannot be determined)

Column 3: Action Codes - What Did You Do?
AC01 - Replace (component completely replaced)
AC02 - Repaired (welded, glued, built up)
AC03 - Cleaned (contamination removed)
AC04 - Adjusted (voltage, pressure, speed modified)
AC05 - Lubricated (grease, oil applied)
AC06 - Tightened (bolts, couplings secured)
AC07 - Reset (system restarted)
AC08 - Temporarily Repaired (emergency repair, temporary solution)

"But what if my problem isn't exactly on the list?" Sandra asked.
"Then we also have free text fields," Pedro explained. "Codes for quick analysis, text for details."

From Chaos to Clarity

Marco studied the lists. "So if I want to replace 'mechanical defect' with 'wear,' I do that in a dropdown menu as well?"
"Exactly!" Pedro replied enthusiastically. "Then we can later search for all the 'wear' problems together and discover patterns."
Jake saw the possibilities. "And if we want to know which problems Marco solves most often?"
"Also possible," Pedro smiled. "Sort by technician, by time, by asset, by cause — whatever you want to analyze."

2.3. Codes as a Filter, Not as a Complete Solution

Pedro grabbed a marker and drew a large funnel on the board. "This is where the real power of coding lies. Imagine we have thousands of failure reports. Without codes, you'd have to read every single one to find patterns."

He pointed to the top of the funnel. "But with codes you filter first. For example: all failures with SC02 - Leak and OC06 - Assembly Error. That might be only twenty instead of a thousand."

"And then?" Sandra asked.

"Then you read those twenty descriptions to investigate what exactly goes wrong during assembly. Maybe you'll discover that it's always the same supplier or that it always happens on Friday afternoon."

Marco nodded. "Like last time with those computers that had no internet connection. If I had been able to filter on 'IT problems' and 'connection,' I would have seen much faster that it always happened after updates."

"Exactly!" Pedro said. "The codes get you quickly to the interesting cases. But the codes don't tell the whole story. 'No internet connection' can still have many different causes. The details in the description tell the real story."

Practical Example

"Let's try it out," Marco suggested. "Yesterday I had a conveyor belt that kept stopping."

"Okay," Pedro said. "What did you see exactly?"

"The belt kept seizing up, making cracking noises."

"That'll be SC08 - Vibrates/Moves. And what was the cause?"

Marco thought about it. "We had postponed maintenance last week, because we were so busy."

"OC03 - Maintenance Backlog. And your action?"

"I lubricated the guide rollers and adjusted the tension."

"Perfect: AC05 - Lubricated and AC04 - Adjusted. You just coded your first failure notification!"

Linda looked surprised. "It's actually quite logical. Much more systematic than our old way."

Additional Categories for Analysis

"There are more ways to categorize," Pedro continued, as he added a new column.

Cost Centers and Work Groups:

- Plant A, B, C
- Warehouse, office, outdoor area
- Electrical, mechanical, civil
- Preventive, corrective, modification

"These extra categories help clarify responsibilities and keep track of budgets," he explained.
Jake got excited. "Then we can finally see how much money we're losing on electrical failures in hall B!"
"Or which technician performs the most replacements," Marco added.
"Exactly. As a reliability engineer, I can then analyze which failures occur most frequently and have the greatest impact," Pedro agreed.

2.4. The Power of Standardization

Sandra was beginning to understand as well. "This is much more practical than those endless lists I've seen from the petroleum industry. Those had codes for things we never encounter."
"That's right," Pedro agreed. "These codes are based on international standards such as ISO 14224, but adapted to our day-to-day practice. The most important thing is that everyone understands the same thing by each code."
Jake looked doubtfully at the lists. "But wait a moment. Not all assets have the same problems? A pump can leak, but an electric motor can't. Shouldn't we have different codes per category? For pumps, ball valves, hydraulic units, presses, mills..."
Pedro grinned. "Good question! That seems logical, but then you get an explosion of codes." He picked up the marker again. "Say we have a hundred categories of assets, each with ten failure codes, each with ten possible causes and five possible action codes..."

He started calculating on the board: "100 × 10 × 10 × 5 = 50,000 different codes!"
Marco whistled. "That's going to be a lot."
"Exactly," Pedro said. "Nobody can memorize 50,000 codes. And who is going to manage and maintain all those codes? That's why companies usually use generic codes that work for all equipment."

"But if we have very critical equipment or failures that occur very frequently," he added, "we can always add specific sub-codes. For example, SC02.1 for 'Hydraulic leakage' or OC06.2 for 'Wrong coupling installed'. The starting point is: start simple, refine where needed."
Linda nodded in understanding. "So we start broad and become more targeted where it truly adds value."
"And if we look at all the SC06 codes in six months," Pedro added, "we'll immediately see which machines wear out the most. Or if we have a lot of OC03, we know we need to plan better."
Marco grew excited. "And with OC06 - Assembly Error, we can look at whether certain technicians need additional training!"

Implementation Step by Step

"Where do we start?" Linda asked practically.
Pedro smiled. "Easy does it. We start small:

Week 1-2: Only use the most common codes - maybe five per category
Week 3-4: Train the team so that everyone understands the same thing by each code
Month 2: Evaluate and Add Codes if Necessary
From month 3 onward: Always enter at least one code from each category

And very important: we always leave room for additional description. The codes are tools, not straitjackets."
Sandra nodded. "And if it turns out we're missing a certain code?"
"Then we'll add those," Pedro said. "This system grows with our experience. In a year, we'll probably have added a few codes and maybe dropped some that we never use."

The Bigger Picture

Linda stood up and looked at the whiteboard. "I'm starting to see it. With these codes, we can finally recognize patterns instead of treating every failure as a separate incident."
"Exactly," Pedro said. "Effective categorization makes it possible to analyze failures in a structured way and discover patterns. Instead of being reactive, we become proactive. We see trends before they become problems."

"And if the EAM system database is filled in correctly with these kinds of codes," he added, "then I can use them as a reliability engineer to pinpoint exactly where we need to focus our energy."
Marco closed his notebook. "When do we start?"
"Starting tomorrow," Pedro smiled. "I'll give you your first training. In two weeks we'll evaluate how it's going."
As the team walked away, Pedro heard Sandra say to Linda: "For the first time in a long while, I feel like we're actually going to bring some structure to that chaos."

2.5. Implementing Quality Assurance

Jake asked how they would ensure that the new registration would remain consistent.
"How do we prevent ending up back in the same old chaos three months from now?"
"By building in quality assurance. Someone needs to review every notification before it becomes final."
James, the work planner who had joined the discussion, volunteered, "I could do that. I already evaluate all work activities on a weekly basis."
"Perfect! You know the EAM system best and see all the notifications."
"What do I need to check?" James asked.
"Completeness, clarity, correct categories. And you can provide feedback on the quality of notifications."
Sandra was curious. "Feedback?"
"If James sees that your reports are always complete, he says so. If they are vague, he asks for clarification."

"So we learn from each other?" Marco asked.

"And I can also see patterns. If I see the same vague notification three times, I know we need training," James added.
Tom nodded. "Or if I forget technical details?"
"Then I'll help remind you of what's important for the analysis."
To ensure the quality of failure information, several measures are essential. Training of reporters ensures that everyone who records failures is well trained in recognizing and describing failures.

Standardized forms help to collect all essential information and provide drop-down lists where possible.
A quality control system ensures that reports are checked for completeness and accuracy before they are finalized.
An important aspect is implementing a feedback mechanism that informs reporters about the quality of their reports and how these can be improved.

2.6. Improving the Registration Process

Sandra complained about the current registration process during one of the following meetings.
"I have to go from the machine to the office to file a report. By the time I get there, I've forgotten half of it."
"We're going to do something about that. We can make the registration process much more effective," Pedro promised. "We can install an app from our EAM on our phones, so that all failure technicians can immediately fill in the most important data on their phone and attach photos. If needed, they can then elaborate on the report later at the computer.

"So I can take a photo of the broken part and attach it to the report later?" Marco asked.
"Exactly. A picture is worth a thousand words. These days you can even dictate the text and don't have to type anymore."
Tom was intrigued. "And that works in the noisy shop floor as well?"

Sandra said, "That would help enormously. Right now I always try to describe what something sounds like or looks like."
Jake asked, "And it just stays in our existing EAM system?"
"Absolutely. We're not replacing the system, we're just making the registration process more user-friendly."
Traditional failure registration was often done manually on paper or through fixed terminals. However, the registration process can be made much more effective through practical improvements.

Direct on-site registration prevents important details from being forgotten between the moment of discovery and the moment of recording. Photos of defects provide far more context than text descriptions alone. A photo of a leak or visual documentation of abnormal wear can significantly improve the diagnosis.

2.7. Organizing Training and Acceptance

Pedro organized training for the new registration system. "Next week we're going to implement the new system. Everyone will receive training."
Marco was practically minded: "How much extra time is this going to cost me?"
"In the beginning maybe five extra minutes per notification. But after a month it should be faster than the old system."
The training consisted of three parts:

- Learning New Categories and Codes
- Practicing with the Improved Forms
- Understanding Quality Criteria and Feedback

"And what if I forget?" Sandra asked. "The forms remind you of required fields. And James gives feedback if something is unclear."
Tom was worried. "What if the old habits come back?"
"Then I'm here to help everyone remember. This isn't a punishment, it's an investment in better problem-solving," Jake assured.
Linda asked, "And we can see the results?"
"In a month I'll show you the first patterns. Then you'll understand why this investment is worth it."

The Medical Analogy

Pedro wanted the team to understand why good record-keeping was so crucial.
Sandra asked, "Why is all of this so important? We just repair what breaks down, right?"
"Think of a doctor. If your primary care physician doesn't keep track of your medical history, how can she recognize patterns in your health?"
Marco nodded. "Doctors have files."
"Exactly! And if she doesn't know which medications you've already tried, how can she prescribe an effective treatment?" Tom said, "So we're doctors for machines?"
"Sort of. By tracking all the 'symptoms' and 'treatments,' we can identify patterns."
Linda said, "And making diagnoses?"
"And above all: prevent the same 'disease' from coming back. If a doctor sees that you get bronchitis every winter, she advises prevention."
Sandra applied it to their situation: "And if we see that section C blocks every month?"
"Then we'll systematically search for the underlying cause and a structural solution."

Jake summed it up well: "So good record-keeping is an investment in the future health of our machines?"
To illustrate the importance of proper failure registration, you can compare it to the medical world. By consistently recording all failures, including symptoms, diagnosis, and repairs performed, you build up a valuable knowledge base.
This knowledge base enables you to recognize patterns, measure the effectiveness of solutions, facilitate knowledge transfer, and make data-driven decisions.

2.8. The First Results

Four weeks later, Pedro organized an evaluation meeting. The team sat around the table while he showed the first results of the new registration system.

"Four weeks ago we started with the new system. Look at what we can see now."
On his laptop, he showed clear patterns in the data.
Sandra said, "Wow, those conveyor belt jams really have a pattern. Especially in the mornings, especially during product changeovers."
Marco said, "And I see that my repairs are lasting longer than those from last month." Tom discovered clusters: "Those electrical problems are clustering around certain assets."
Jake asked, "And the extra time it took?"
"Already paid for itself. Notifications are much clearer, I have to follow up less," James replied. "And this is just the beginning. Now that we have good data, we can set priorities and do real Root Cause Analysis," Pedro added. Sandra admitted, "I have to admit, it feels good to see patterns instead of just reacting."
Pedro smiled at his team. "Next week we'll start holding daily meetings to separate acute problems from structural challenges."

Through systematic registration, you can make better predictions about future failures and optimize maintenance cycles. With the information in this database, you as a reliability engineer can analyze what the most significant production losses and risks are.
Based on that, you can set priorities and determine the root causes of the problems. Modern EAM systems offer various functionalities that support the failure registration process, but the foundation always remains good, systematic data collection.
Pedro looked with satisfaction at the results of their efforts. His laptop no longer showed chaos, but clear patterns. The team had learned to describe problems systematically instead of randomly noting symptoms.

The conveyor belt line three was still their biggest challenge with 78 jams per month, but now they finally had the right data to tackle the problem. Sandra's observation about the timing of the jams was valuable. Marco's technical details provided insight into the mechanical aspects. Tom's electrical measurements added a new dimension.
Now that they had set up a solid registration system that collected the right data, they could begin managing those failure reports on a daily

basis. In chapter 3 you will read how daily meetings make an initial triage between acute problems and structural improvements — the way to respond quickly to acute problems and distinguish between minor issues that can be resolved immediately and larger problems that require a more structural approach.

3. Daily Failure Review

3.1. The Daily Failure Review Meeting

Now that you have set up a solid registration system that collects the right data, you can begin managing those failure notifications on a daily basis. The next crucial step is regularly discussing these failures through a daily failure review meeting.

Pedro's First Attempt

At InduPack Solutions, reliability engineer Pedro stood in the conference room on Tuesday morning at a quarter to eight. On the whiteboard he had written the agenda for their first official daily failure review meeting. The monitor displayed the failure data from the past 24 hours.

At exactly eight o'clock, the team members trickled in. Sandra arrived first, with a steaming coffee mug and a skeptical look.
Jake, the production supervisor, glanced at his watch. "Management wants us to spend less time in meetings and more time on repairs."
Pedro set down his laptop and looked around at the team. "I understand the resistance. But give me three weeks. If this doesn't add value, we'll stop."
Tom, the electrical technician, came in last. "How long is this going to take?"
"Twenty minutes max," Pedro replied. "And we focus on facts and solutions, not on who did something wrong."
Linda nodded toward the screen. "Yesterday, four jams on line three. Total of 48 minutes of production loss."
"Exactly," Pedro said. "Facts first, then patterns, then actions. Let's get started."

Sandra looked at the agenda on the whiteboard. "And we're not going to look for someone to blame?"
"Absolutely not," Pedro replied. "We're looking for patterns and solutions. It's about improving systems, not punishing people."

Discovering Structure

Pedro walked to the whiteboard and pointed to the agenda he had prepared:

Daily agenda:

- Summary from yesterday: Key failures, impact, current status
- Immediate actions: What needs to happen today, who does what
- Patterns: Do we see repetitions, trends
- Structural improvement: Are there improvements that we can already see and easily implement

"I suggest we maintain the same rhythm every day," he said. "That keeps us focused."
Marco frowned. "What if there was nothing unusual yesterday?"
"'Nothing unusual' is also information," Pedro said. "Maybe the circumstances were different. What can we learn from that?"
Jake nodded slowly. "And if there had been a major failure?"

"Then we discuss the immediate actions to resolve it, but we also ask: is this an incident or a symptom of a larger problem?"
Tom leaned forward. "How do we know that difference?"
"By looking at it systematically," Pedro replied. "First time? Probably an incident. Third time? Probably a pattern."
Linda made a note. "And patterns mean we need to dig deeper?"
"Exactly. That's why we keep every meeting short and focused."

3.2. The First Breakthrough

After a week of daily discussions, Sandra brought up an interesting observation: “That conveyor belt jam from yesterday... actually began a few days ago.”

"What do you mean?" Pedro asked.
Sandra leaned forward. "Monday I had to adjust that guide roller once per shift. Tuesday it was twice. Wednesday three times. Yesterday it really

went wrong — three times per hour — that's when we reported it as a failure."
Tom nodded understandingly. "So it gradually got worse."
"Exactly," Sandra said. "At first I thought: just make a quick adjustment. But now that I think about it..." Pedro wrote on the whiteboard:

Day 1: 1x per shift →
Day 2: 2x per shift →
Day 3: 3x per shift →
Day 4: 3x per hour = FAILURE

"This is worth its weight in gold," Pedro said enthusiastically. "We definitely see this pattern more often."

Jake nodded. "Things gradually come loose, wear out, or get thrown off. Operators compensate for that initially."
Marco frowned. "And by the time they call us in, it's already a major failure. And in the meantime, we've already lost a lot of production."
"Whereas," Pedro added, "if we had known this three days earlier, Marco might have simply been able to tighten that roll."
Linda made a note. "So we can discuss these kinds of signals here?"
"Absolutely," Pedro replied. "You notice gradual changes that we technicians miss. By sharing it here, we can intervene before it escalates."

Sandra looked at the whiteboard. "I hadn't realized that those small annoyances were important."
"They are actually very valuable," Pedro said. "Early signals give us the chance to prevent problems instead of repairing them."

Puzzle Pieces of Information

"That conveyor belt jammed," Tom said during the next meeting. "I remember adjusting the tension on that roller last month after a major maintenance overhaul."
Marco looked up. "Wait, which maintenance service?"
"The one on the main drive," Tom replied. "We took everything apart at that point."

Linda frowned. "And was the roller realigned after that?"
"I'm not sure," Tom admitted. "I only adjusted the tension."
Pedro noted this on the whiteboard:

Possible cause: Misalignment following maintenance service last month

"This is exactly what it's all about," he said. "Tom, without this discussion, this information would have been lost."
Jake nodded. "And by the time we do a major analysis, we may have forgotten those details."
"Or the log files have been overwritten," Linda added.
Sandra looked at her colleagues. "I hadn't expected that you also had information about my problem."

"That's why this is so valuable," Pedro explained. "Production sees the symptoms, maintenance knows the history, engineering knows the technical details. By bringing this together, we get a more complete picture."
Marco made a note. "And even if we don't prioritize this problem right away, we'll soon have all the puzzle pieces."
"Exactly," Pedro agreed. "Now while everything is still fresh in everyone's memory and in the systems."

3.3. Creating Ownership

Jake led the discussion during the Friday meeting. "How do we make sure actions actually get done?"
"Good question," Pedro replied. "I think: one owner per action, clear goals, and follow-up."
"What do you mean?" Marco asked.
Pedro stood up and wrote on the whiteboard:

Marco checks tension
versus
Maintenance checks tension

"Do you see the difference?" Pedro asked. "In the first case, Marco knows exactly what is expected of him."
"And make it specific," Jake added. "Not just 'what' but also 'when.'"
Pedro nodded and wrote an example:

Problem:
Conveyor belt jam section C
Acute action:
Marco checks the tension of guide roller 7 before 3:00 PM

Sandra asked, "And if Marco doesn't make it by 3:00 PM?"
Marco said, "Then I'll report why not and when I will."
"Exactly," Pedro said. "Transparency is important."
Jake picked up a marker. "I suggest that I update the action board every morning. That way everyone can see what's still open. We no longer erase this board. The information stays up all day. Only when we decide in the morning meeting that we're done with an item, then it can come off the board."

Tom nodded. "And we'll check the status at the next meeting."
"Ownership and follow-up," Pedro concluded. "That's how we make sure actions don't fall through the cracks."

Seeing Patterns

After three weeks, Pedro had drawn a simple graph on the whiteboard. "Three weeks of data. Take a look at this line."
The chart showed failures per day, time of day, and product type.
Linda studied the numbers. "Clear peak between 6:00–8:00 AM."
Tom nodded. "And especially with dark paper."
Sandra smiled. "I see that on the shop floor too, but now we have proof."
Marco pointed to another part of the graph. "And it's worse on warm days?"
"That's why we visualize it," Pedro explained. "Patterns that aren't visible day-to day become clear over time."
Jake leaned back. "We could look at different types of patterns."
"Such as?" Linda asked.

Pedro began writing on the whiteboard:

Time patterns - time of day, day of week. Environmental patterns - weather, temperature.
Process patterns - product changeovers, startups.

"And human patterns," Sandra added. "Different results per shift."
"And asset patterns," Marco said. "Similar problems on identical machines."
"Fantastic," Pedro said. "You're already thinking systematically."
Marco looked thoughtfully at the graph. "Those conveyor belt jams between 2:00 and 4:00 PM... I think I know why that is."
"Tell me," Pedro said.

“That guide roller is positioned just a little too far to the right. Small deviation, but at elevated temperatures the material expands and then the problem gets worse. If I shift that roller two millimeters to the left, that might fix it. Takes me ten minutes.”

Sandra looked surprised. "Would that really help?"
"The theory checks out," Tom said. "Expansion from heat."
Pedro smiled broadly. "This is exactly what I mean. Marco, can you do that today?"
"No problem," Marco replied. "Before lunch."
Jake made a note. "And then we monitor whether it helps."
Pedro leaned forward. "This is an important point. We distinguish between two types of problems in these discussions.
First: problems where the solution is clear and takes little time — like shifting Marco’s roller. We address those directly.

Second: complex problems that require thorough analysis. We won’t discuss those in detail here. Those will be prioritized in our follow-up process.”

Linda nodded understandingly. "So no lengthy technical discussions about complicated problems."

"Exactly," Pedro agreed. "We want quick results without bureaucracy. A simple adjustment that solves a problem, we just do it. We save complex analyses for later."
Tom grinned. "Quick wins versus deep dive."

"Well said," Pedro laughed. "This meeting is for the quick wins and gathering information. The deep dives will come up when we set priorities."
Sandra looked at the graph. "Now I see why those daily meetings are valuable."
Marco nodded enthusiastically. “We’re building up knowledge about why things happen, and we can act quickly when the solution is obvious.”
"And we can document that," Linda said. "A simple log of what we observe."

Melting Resistance

After two weeks of daily meetings, Sandra voiced her doubts. "I still don't see what these meetings are getting us. We're talking about problems I already know about."
Pedro leaned forward. "What do you know about those conveyor belt jams?"
"They happen mainly in the morning. Usually during product changeovers. Especially with coated paper," Sandra replied.
Marco looked up in surprise. "I didn't know that. I thought it was just random."
Tom frowned. "And I thought it was mechanical. Not related to paper type."
Linda made a note. "Sandra, you see patterns that we miss, because you work with it every day."
Jake nodded. "And by discussing it, we share that knowledge."
"Exactly!" Pedro said. "Your experience becomes team knowledge. And team knowledge leads to better solutions."
Marco added: "And I hadn't realized that timing was so important."
"That's why these discussions work," Pedro concluded. "Not for new information, but for shared information."
Sandra looked thoughtful. "I'm starting to see it. It's not about what I know, but about what we know together."

"And that's much more than the sum of its parts," Pedro agreed.

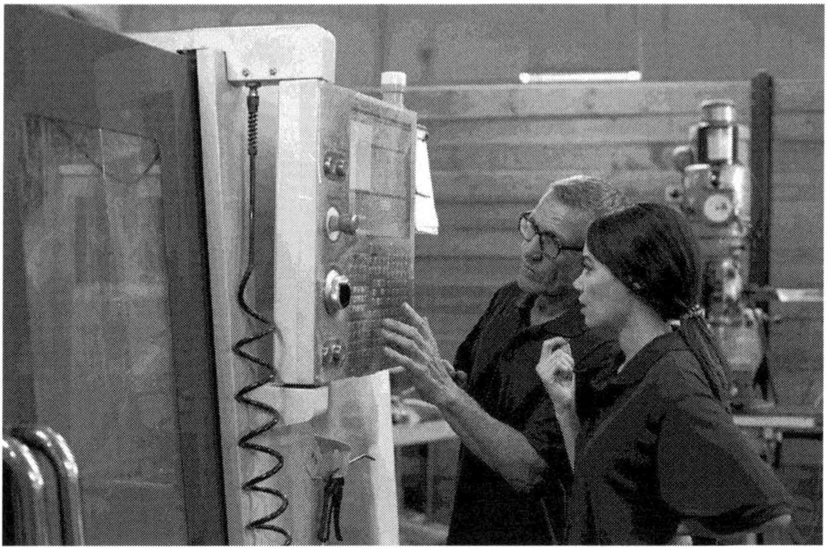

The Culture Shift

Jake reflected during one of the later discussions. "We used to be firefighters. Running from failure to failure."

"And now?" Pedro asked.

"Now we're detectives. We're looking for patterns and underlying causes."

Sandra nodded. "It feels different. Less hectic."

Marco added: "I still repair things, but now I understand the context. And sometimes I can prevent problems."

Tom smiled. "And we share knowledge. I now know things about other lines that help me with my own work."

Linda picked up her notebook. "I've been tracking some numbers from this month versus last month."

She wrote on the whiteboard:

Previous month:	78 reported failures
This month:	52 reported failures, but 23 problems caught early

"We have fewer official failures," Pedro said, "but we're solving more problems."
Sandra added: "And the beautiful thing is that operators now dare to report when something 'isn't quite running right,' even if it isn't a failure yet."
"That's the biggest gain," Pedro agreed. "You are our early warning system."

3.4. The Results Speak

Six weeks after the start, Pedro wanted to share the results with the team. He had put some numbers on the whiteboard.
"Six weeks ago we started with daily failure reviews. You were skeptical," he began.
Sandra laughed. "I thought: 'Another pointless meeting.'"
"And now?" Pedro pointed to the numbers on the whiteboard:

Failure frequency:	-33% (78 → 52 per month)
Mean repair time:	-28% (45 → 32 minutes)
Recurring failures:	-45%
Early detected problems:	+23 per month

Jake whistled admiringly. "The numbers speak for themselves."
Linda nodded. "But more importantly: we have identified eight different patterns."
Marco was enthusiastic. “And when I do a repair, I think: ‘Is this part of a larger pattern?’”

Avoiding Pitfalls

Over the course of the weeks, the team discovered what worked and what didn't.

"What have we learned from these discussions?" Pedro asked during an evaluation.
Jake began: "Stick to the schedule. Nothing kills enthusiasm faster than meetings that run over."
Sandra nodded. “And only discuss the main points. We can go into the details later in more depth.”

Marco added: “Make sure every action has an owner and a deadline. Otherwise nothing will happen.”

Tom grinned. "And make sure there are always substitutes when someone is away. Continuity is important."
Linda looked up from her notes. "And celebrate the successes. It helps to see that it works."
"Important lessons," Pedro agreed. "And the most important one?"
"Stick with it," Jake said. "Even when the first results disappoint. It takes time to take hold."

The Road Ahead

Jake leaned forward. "What's the next step?"
"We now have a good system to collect information and pick up early signals," Pedro replied. "The next step is to determine which problems we're going to investigate more deeply."
Sandra looked at their collected data on the whiteboard. "We've gathered quite a lot of information over these weeks."
"Exactly. And we can't tackle everything at once," Pedro said. "We need to set priorities."
Marco nodded. "I have confidence in it. This systematic approach works."
"And that's just the beginning," Pedro said.

Tom looked at his colleagues. "I'm curious what we'll have achieved in a few months."
"We'll find out together," Pedro replied.

In chapter 4, we follow Pedro and his team as they learn how to prioritize which problems to investigate more thoroughly. Because, although they have now collected a wealth of information, they cannot tackle everything at once. How do you determine what is most important? And how do you ensure that you can make the greatest impact with your limited time and resources?

4. Setting Priorities

4.1. Criteria for Setting Priorities

The conference room of InduPack Solutions was covered in whiteboards covered with patterns, trends, and failure categories. Pedro stood before a list of 23 different problems they had identified over the past weeks. Each problem seemed important enough for a structural approach.
Sandra the experienced operator of line three, looked overwhelmed at the list. "23 problems! Where on earth do we start?"
Jake Jensen, the production supervisor, felt the pressure from management. "They're asking for quick results. They want us to address all the problems."
Marco Rivera, the mechanical technician, shook his head. "I can't work on 23 projects at the same time. That's going to be chaos."

Pedro nodded understandingly. "That's exactly the problem I was wrestling with at TechCorp. We were trying to tackle 127 problems there all at once."
Linda Wang, the quality engineer, looked up in surprise. “127? How did that turn out?”
"Disastrous," Pedro sighed. "After a year, eight problems had been addressed, three of which didn't work. And those eight were also the wrong priorities."
Tom Davis, the electrical engineer, understood the problem. "So we have to choose?"

"We need to choose wisely," Pedro emphasized. "Focus on the vital few that have the greatest impact."
Sandra frowned. "But how do we determine what 'vital' is? It seems like every problem is important to someone."
"By systematically prioritizing with data," Pedro replied, walking toward the whiteboard. "It's a utopia to think that you can immediately and thoroughly analyze every failure and implement a definitive solution. You have

to choose which problems you're going to work on, and therefore set priorities."
Jake nodded. "And colleagues and management will appreciate that?"
"Often they'll appreciate that you're tackling the most important problems first," Pedro replied. "What they often find annoying is that you're not addressing certain less important issues. Yet it is important to make this very clear to everyone."

The Priority Decision Framework

Sandra looked at the list of 23 problems. "How do we determine which problems deserve extensive analysis?" Pedro shook his head. "By first applying a practical filter. Not every problem requires the same approach."
He drew a diagram on the whiteboard with two axes: Impact and Investment.
"For every failure, you ask yourself two questions," he explained. "What is the impact? And how much time and money does a solution cost?"
He filled in four quadrants:

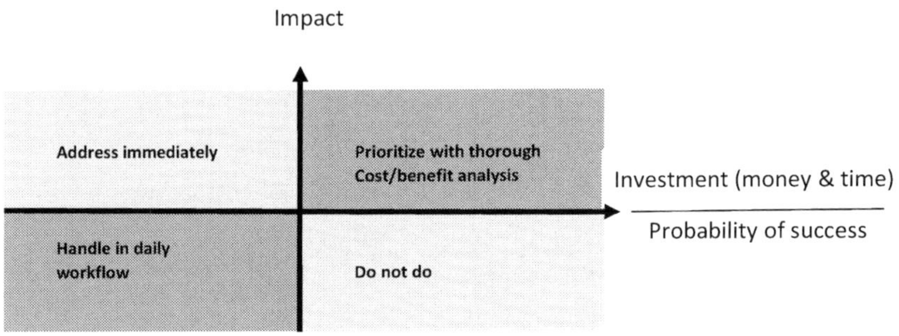

Figure 4-1 Decision Matrix for Problem Approach

- High impact, low investment, and a high probability of success: "You tackle these immediately. A technician can solve it in an hour, or a small team action resolves it right away."
- Low impact, low investment, and with a high probability of success: "Make it a regular work order. No project, just handle it in the daily workflow."

- High impact, high investment, or with a low probability of success: "These require systematic prioritization with thorough analysis. We can't tackle all the problems at once here."
- Low impact, high investment, or with a small chance of success: "You deliberately choose not to address these. Say out loud that you're not doing anything about them, otherwise they'll keep coming back in meetings.

Marco nodded understandingly. "So we don't need to analyze everything extensively?"
"Exactly," Pedro agreed. "Save your energy for the problems that are worth it."
Jake saw the practical value. "And for our 23 problems?"
“Let’s run them through this framework first,” Pedro suggested. “Then we know which problems require Pareto analysis.”

4.2. The Danger of Lack of Focus

Pedro leaned against the whiteboard and talked about his own learning experiences. "When I started out as a young engineer, I made exactly the same mistake."
He described how the production manager had assigned him a problem in his first week. Enthusiastically, he had gotten to work on it. The second week he received two more. The third week even more.
"By the fourth week I had already come up with a solution for the first problem," Pedro said. "I had consulted with the supplier, made a plan, arranged the budget. But it took months to actually implement."

Jake nodded in recognition. "And in the meantime, new problems kept coming up?"
"Exactly. After five weeks my manager said: 'Let that week-one problem rest, it hasn't occurred again.' So I set it aside and started on a new problem."
Sandra understood where this was going. "And a year later?"

"The first problem came up again. 'Why hasn't anything been done about this?' management asked then. They had forgotten that they had told me to let it rest."
Marco shook his head. "Vicious cycle."
"That's why," Pedro said firmly, "we work in a focused way here. A limited number of problems, handled completely, then move on to the next one."
Tom asked, "What do you mean by fully resolving it?"
"Not only the technical implementation," Pedro explained, "but also updating documentation, training operators and technicians, and evaluating the solution. Only after you have determined that no solution is possible, or after you have truly solved a problem, can you move on to the next one."

4.3. Chronic versus Sporadic Failures

Linda studied the list on the whiteboard. "Some of the problems here are minor irritations we constantly deal with, others are major disasters that happen occasionally. How do we compare those?"
Pedro picked up a clean marker. "By distinguishing between chronic and sporadic failures."

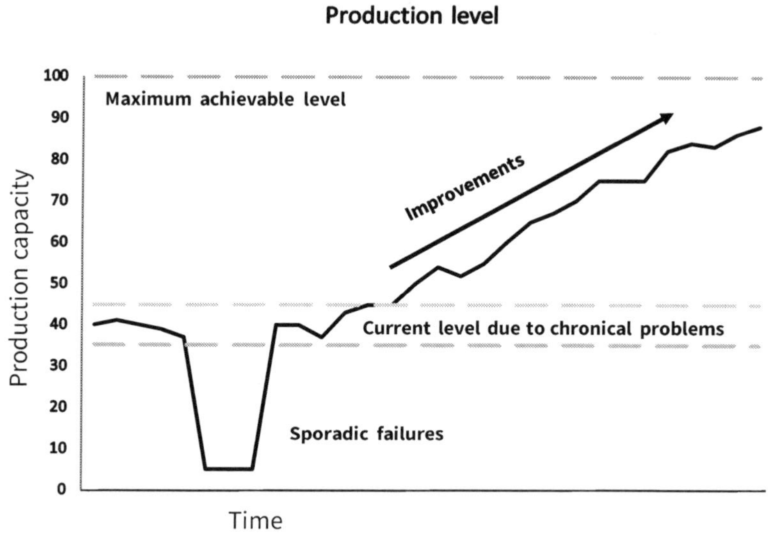

Figure 4-2 Impact of chronic and sporadic failures

He began drawing a diagram. “Chronic failures,” he explained, as he wrote, “have a high frequency but low impact per incident. They are often considered ‘normal,’ because we have gotten used to them.”

Sandra looked up. "Those small jams I continuously solve - those are chronic failures?"
"Exactly. You resolve them so quickly that nobody notices. But together they may cost more than the major failure that everyone remembers. If you lose five minutes every hour, that adds up to fourteen hours per week."
Marco thought about a recent incident. "And that main bearing that failed last year?"
Tom added: "Sporadically. Once a year, but two days of downtime."

Jake asked the obvious. "Which one takes priority?"
"We determine that with data, not with gut feeling," Pedro replied. "Both types can be equally important."
Linda nodded understandingly. "So we need to include both types in our analysis?"
"Absolutely," Pedro agreed. "A great tool for that is constantly measuring the Overall Equipment Effectiveness - the OEE."

4.4. Overall Equipment Effectiveness Monitoring

Sandra frowned. "OEE? What exactly is that?"
Pedro wrote on the whiteboard, "OEE is the percentage of theoretical production capacity that your company actually achieves."

He gave an example: "If the production line of a soft drink factory can produce 10,000 bottles per hour and the production staff is present 120 hours per week, then the theoretical production capacity is 1,200,000 bottles per week. If they are then down 12.5% of the time, running at only 90% of their capacity during the time they are producing, and also have 5% rejects, they produce fewer than 900,000 bottles per week."
Tom did the math: "But 897,750 to be exact. That's only 74.8% of what they could be making?"

"Then the OEE that week was 74.8%," Pedro confirmed. "An OEE measurement provides insight into where the other 25% was lost."
Marco was curious: "How do we categorize those losses?"
Pedro drew a diagram. "A commonly used classification is to divide the losses into six categories:"
He wrote, while explaining:

"Downtime losses: planned downtime such as changeovers or preventive maintenance, and unplanned downtime such as failures.

Speed losses: slower line running and minor losses.

Quality losses: rejected products at startup and random rejects during production."

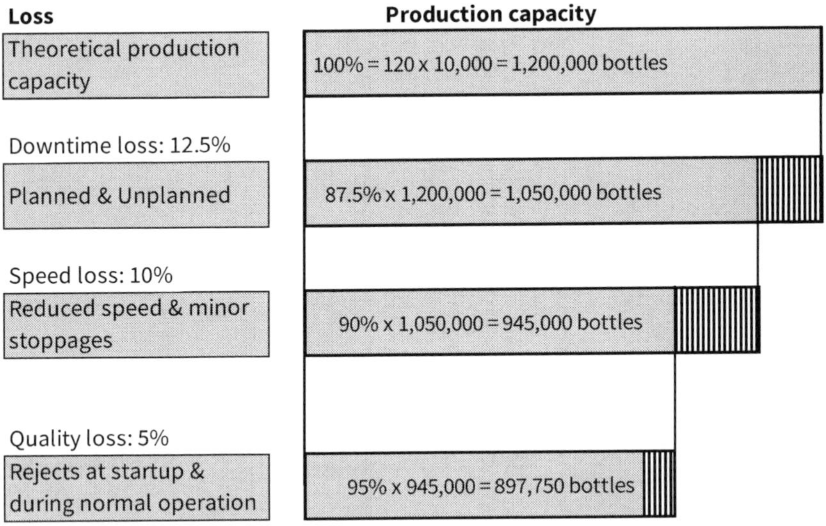

Figure 4-3 The six losses of Overall Equipment Effectiveness

Linda saw the value immediately. "By systematically measuring your OEE, you gain insight into chronic failures that would otherwise remain invisible?"
"Exactly," Pedro confirmed. "Those small losses that Sandra keeps solving are now visible in the data."
Jake looked at the diagram. "And sporadic failures?"

"Those are easier to recognize," Pedro replied. "Sporadic failures are rare but often major events with significant impact. They occur infrequently, maybe once a month or a year, but each failure causes a lot of downtime."
Sandra nodded. "Those have high visibility and attract management's attention."
"That's right," Pedro said. "These failures are almost always recorded in the EAM system and are often more complex to resolve. One example is a main bearing that fails in a critical production line, leading to days of downtime and high repair costs."
Tom summarized: "By systematically addressing both chronic and sporadic failures, we can significantly reduce total failure losses?"
"That's exactly the point," Pedro agreed.

Pedro's Priority Breakthrough

"Six weeks ago I was standing where you are now," Pedro said, while he wrote data on the whiteboard. "Overwhelmed by choices at conveyor belt line three."
Jake leaned forward. "What did you do then?"
"I ran a Pareto analysis on three months of conveyor belt data." Pedro drew a chart on the board. He showed the results he had found:

- 78 different jam types identified
- Top 4 types cause 73% of loss
- Other 74 types combined: 27% of loss

Sandra let out a soft whistle. "Four out of 78? That's only 5%!"
"The 80/20 principle in action," Pedro confirmed. "20% of causes, 80% of impact."
Linda was curious. "What were those four types?"

- Paper jam section C during product changeovers
- Sensor errors at low humidity
- Mechanical lockout after weekend shutdown
- Electrical failure during startup

Tom understood the implication. "And instead of addressing all 78..."

"Are we going to focus on those four," Pedro completed. "These four chronic jam problems were so small that operators resolved them quickly without much attention. But together they caused $180,000 of the total $240,000 in annual losses."
Jake saw the logic. "So you decided to focus all your energy on these four priorities?"
"Instead of all 78 problems, exactly."

4.5. Creating Pareto Charts

Linda stood up and walked to the whiteboard. "Let's do the same for all our problems. I can analyze the data."
Pedro nodded enthusiastically. "A Pareto chart is a powerful tool for identifying the most important causes of failures."
He began to draw. “A Pareto chart is a bar chart where each bar on the X-axis represents a cause of losses. The height of the bar shows the magnitude of the losses caused by that factor.”

Sandra looked at the diagram. "And how do you rank those bars?"
“The tallest bar is on the left, against the Y-axis. The second-largest problem is next to it, and so on. This way you can see at a glance what the most important problems are.”

Tom asked practically: "How do we make such a diagram?"
Pedro explained the steps: "Start by collecting failure reports from your EAM system. Make sure you have at least three months of data for reliable results. Then group similar failures into logical categories."
Marco followed the reasoning. "All conveyor belt jams combined?"
"Exactly. For each category, you calculate the total impact. This can be in time, money, energy, or other relevant units."
Linda took notes. "And then rank them from high to low?"
"Yes, and an important step is calculating cumulative percentages. Add a line that shows the cumulative percentage of the total impact. This helps identify the 'vital few.'"
Jake understood the concept. "The losses on the Y-axis can be different types?"

"That's right," Pedro confirmed. "Downtime, frequency of failures, energy loss, rejects, costs — depending on what matters to your company."
Sandra saw the value. "It's a powerful visual tool. Everyone can immediately see which failures are responsible for the vast majority of the losses."

Practical Example from the Dairy Industry

Pedro shared an example: "In a dairy plant, we created a Pareto chart of the number of downtime minutes caused by failures per machine. This showed that 20% of the machines were responsible for 76% of the downtime."
Tom was interested. "And then?"
"It also turned out that the two machines that caused the most failures were in fact identical. We then pulled the failures for those machines from the failure database and thoroughly examined which recurring failures were occurring."

Linda anticipated: "And another Pareto pattern?"

Figure 4-4 A Pareto chart

"Exactly. 78% of the downtime on these machines was caused by 20% of the causes. Our analysis showed that the leakage of a number of cooling hoses on these two machines led to a quarter of all the downtime in the entire plant."
Sandra let out a soft whistle. "Two machines, one type of problem, a quarter of all downtime."
"That is the power of systematic analysis," Pedro emphasized.

4.6. Integration of the Risk Matrix

Tom looked at the Pareto results and frowned. "What if a failure rarely occurs, but is life-threatening? That one isn't in our top 4, but..."
Pedro interrupted him with a nod. "Good question. Pareto works for frequency-based problems. For safety, we use a risk matrix."

He drew a new table on the whiteboard and filled in the boxes. Vertically were categories describing how severe the consequence of failure was. This severity was described for specific types of risks.
Pedro described 'Economy, Safety, Quality, Energy Consumption, and Environment'. For each category he defined five types of consequences, which he ranked from 1 through 5. Horizontally alongside that was the frequency at which a particular failure mode could occur. In the matrix, the cells were colored green, yellow, and red.

He gave examples:
"Conveyor belt jams: High probability, Medium impact = High priority. Electric shock: Low probability, High impact = Medium priority. Minor leak: Medium probability, Low impact = Low priority."
"So with this we can weigh safety risks against, for example, environmental risks?" Sandra asked.
Pedro confirmed, "Exactly. We combine Pareto with risk assessment for complete prioritization."
Linda thought practically: "Financial impact and safety risk."
Marco asked the practical question: "How do we weigh those against each other?"

Score	Economics	Safety	Condition	Energy consumption	Environment	Heard of it	<1x per 10 years	<1x per year	<1x per month	>1x per month
1	< $1,100	No impact	No impact	Optimal	No consequences					
2	< $5,500	First aid	Minor complaint	< 5% increase	Internal spill					
3	< $55,000	Lost-time accident	Return delivery to customer	< 10% increase	Reportable incident					
4	< $550,000	Permanent disability		> 10% increase	Off-site damage					
5	> $550,000	Fatality	Product recall		Permanent off-site damage					

Risks in this category require immediate action

Risks in this category require planned action. These must be weighed against each other and prioritized

Risks in this category are acceptable

Figure 4-5 A risk matrix

"In consultation with management," Pedro replied. "They determine the risk level that is acceptable."
Tom summarized: "So some rare failures can receive higher priority than frequent failures with limited consequences?"
"Depending on what the probability is that they occur," Pedro confirmed.

Decision-Making Process

Jake organized a management meeting for the final prioritization. In the boardroom, representatives from production, maintenance, quality, and management gathered.

Jake presented the Pareto results to all stakeholders. The production manager nodded in recognition. "The conveyor belt is at the top. That matches my experience."
The quality manager had other concerns. "But those sensor failures affect our certification."

The maintenance manager pointed to safety issues. "And a failure of the line 2 pump has safety risks."
Jake explained the approach: "That's why we combine financial impact with risk assessment."
The production manager saw the logic. "Conveyor belt wins on total score."
The safety manager remained concerned. "But pump failure has the highest safety score."
Pedro suggested a compromise: "We can tackle both. Two problems at the same time is no problem. Five isn't either, but we need to make sure it doesn't become twenty-five. Then we'll ultimately get nothing done."

Resource-Based Prioritization

Marco looked anxiously at the implementation plan. "How many improvement projects can we realistically handle?"
Pedro thought about it. "It's not about the number of people working on one problem. It's about not starting a structural improvement project every time a failure occurs."
He explained, "Employees can each work on different problems. You, for example, can contribute to five different improvement projects. But if you immediately kick off a major improvement project every time a new failure occurs, you'll quickly end up working on a hundred projects simultaneously and nothing will ever get finished."
"So it's about the number of structural improvement projects, not about daily repairs?" Sandra asked.
"Exactly," Pedro confirmed. "Daily repairs just keep going. But for structural improvements, you deliberately choose a limited number of priorities."
Tom asked, "How much is realistic?"
"That depends on the size of your team and the complexity of the problems," Pedro replied. "I would say: start with one or two priorities, see how that goes, and then expand."
Jake saw the practical side. "And how do we divide the time?"
Pedro wrote an example on the whiteboard:

Conveyor belt project (first priority):

- Pedro: project management and analysis
- Sandra: practical knowledge and testing
- Marco: mechanical solutions
- Tom: electrical aspects

Pump failure project (second priority):

- Marco: lead role mechanical
- Tom: safety systems
- Pedro: facilitation and method

Linda asked about planning: "And timing?"
"That's hard to say in advance," Pedro admitted. "The planning depends on what you find during the investigation. Sometimes a problem is simpler than expected, sometimes more complex."
Marco nodded in understanding. "You often don't know yet what the exact cause is."
"That's right," Pedro confirmed. "And so you only know after an investigation what's needed to solve the problem."
Tom summed up: "So we start with realistic expectations and adjust where necessary?"
"Exactly. Focus, flexibility, and fully resolving things before you move on."

4.7. The Importance of Quick Action on First Results

Sandra could not hide her impatience. "When do we start with the conveyor belt analysis?"
Pedro smiled. "Next week. The faster we show results, the more people will get excited."
Jake nodded in agreement. "Management wants quick, visible progress too."
Marco rubbed his hands together. "And I finally want to solve that problem structurally."

Pedro explained why swift action was important: "First, we maintain momentum. We convert team enthusiasm into concrete action. Second, we

can prove that a systematic approach works. Third, we motivate others to join in. And fourth, we gain practical experience for future projects."
Tom asked practically: "And the first results?"
"It varies by project," Pedro replied honestly. "But within a few weeks to months, we should start seeing initial improvements."
Sandra beamed: "I can't wait to see those jams resolved for good."
Jake saw the broader impact: "And other departments will see our results and want to work systematically too."
Pedro wrapped up optimistically: "From one successful improvement to a culture of systematic problem-solving."

4.8. Verification of Improvements

Tom looked at the numbers on the whiteboard. "How do we know if our approach is actually working?"
Pedro smiled. "Every six weeks we review the Pareto figures from the last six weeks separately and cumulatively over the last two years."
He pointed to the diagram. "When a problem is solved or significantly reduced, it will disappear from the top of the Pareto chart."
Sandra nodded understandingly. "And if it stays at the top anyway?"

"Then we'll discuss whether we see any possibilities to resolve it," Pedro replied.
Jake added: "Or it comes back after a while."
Marco was practical: "So the Pareto analysis tells us whether we are successful."

Real-World Example: Setting Priorities in Action

Pedro shared another example from his experience: "At a metal processing company, there had been various production failures for years. A new maintenance engineer was appointed to implement improvements."
Linda was curious: "What did he do?"
"Instead of jumping into action right away, he began by systematically collecting failure data over a period of six weeks. After this period, he created a Pareto chart that clearly showed that 75% of the production losses were caused by failures on just two machines in the production line."
Tom asked, "And then?"

"Instead of trying to address all failures at once, as had been attempted before, he established a multidisciplinary team. This team focused entirely on these two 'problem systems'."
Marco could see where this was going: "And because of that focus..."
"Could the team deeply analyze the underlying causes of the problems," Pedro completed. "They discovered that both machines suffered from the same fundamental problem: incorrect pressure settings leading to increased wear on certain components."
Sandra was interested in the solution: "And then?"
"After implementing an improved alignment procedure and introducing preventive maintenance at the critical points, the number of failures decreased by 68%."

Jake whistled admiringly: "Within what timeframe?"
"These results were achieved within six months and led to a 12% increase in production, resulting in additional annual revenues of $280,000."
Linda saw the broader lesson: "This success highlighted the importance of targeted prioritization?"

"Instead of spreading efforts across too many different problems at once," Pedro confirmed. "And it convinced all stakeholders of the value of a structured approach and data-driven decision-making."

Tom quoted: "'In the past, I felt like a firefighter constantly running from one fire to the next,' the maintenance engineer said later."
Pedro nodded: "'Now we're acting as a prevention team that systematically identifies and addresses the greatest risks before they lead to problems.' This transformation illustrates well how systematic prioritization leads to structural improvements instead of treating symptoms."

The Road to Problem Definition

Pedro stood in front of the whiteboard with the final priority list. "We have our priorities clear. Conveyor belt line three gets full focus."
Tom asked about the next step: "What are we going to do now?"
"Sharp problem definition," Pedro replied. "From 'conveyor belt has jams' to a specific, analyzable problem."
Sandra asked for clarification: "How specific?"
"What type of jam, under what circumstances, on which part of the machine, during which activities."

Marco understood: "From vague to precise?"
"From unsolvable to solvable. A good problem definition is already half the solution."

Linda saw the connection: "And then systematically collect information?"
"Exactly. All relevant data about that specific problem. Then we're ready for a real Root Cause Analysis."
Tom looked ahead: "Next week we'll start the systematic analysis of conveyor belt line three."
"The real journey begins now," Pedro concluded.
Now that you have learned how to effectively prioritize with Pareto analysis and risk matrix, you are ready for the next step. In chapter 5, you will learn how to sharply define a specifically chosen problem and gather the right information for thorough analysis.

This forms the basis for finding a structural solution through systematic Root Cause Analysis. You will learn how to move from a broad priority (such as "conveyor belt jams") to a sharply defined problem that you can effectively analyze.

5. Problem Definition and Information Gathering

5.1. The Improvement Team

Pedro stood in front of the monitor in the conference room, which displayed a Pareto chart. Around the table sat Sandra, Marco Rivera, Tom, Linda, and Jake Jensen — the team he had assembled for their first structured improvement project.

"Okay team," Pedro began, pointing to the chart, "the Pareto analysis is done. Conveyor line three is our biggest pain point — 34% of all production loss comes from there. But now comes the hard question: what does 'conveyor has problems' actually mean?"
Sandra sighed audibly. After fifteen years as an operator, she had heard this story a hundred times. "I've heard that about a thousand times lately. Every time we say 'the belt fails' as if it's one thing."

Marco nodded in agreement. "Sandra is right. Sometimes it's a paper jam, sometimes a sensor error, sometimes mechanical. We're talking about different problems as if they're the same thing."

Linda looked up from her laptop. "I've gone through the failure reports from the last six months. We use 47 different descriptions for what are essentially 'conveyor belt problems.' From 'belt gets stuck' to 'production interruption section C' to 'jam zone 3.' It's completely chaotic. Luckily, our new failure report is going to solve this in the future."
Pedro smiled. This was exactly what he had hoped to hear. "Exactly! And that's why today we're going from 'the conveyor has problems' to a sharp problem definition. From vagueness to specificity. Because you can't solve what you can't define."

Sandra's Gold Mine of Practical Knowledge

Pedro turned to Sandra. "Sandra, you're the one who deals with those jams every day. Can you tell us exactly what different types you see?"

Sandra leaned back in her chair. She wasn't used to someone taking her expertise so seriously. "Look, there are actually four main types of jams. The paper jam in section C — that's the worst one. Then you have the sensor fault at the output, the mechanical blockage at the input, and those strange electrical stops that nobody understands."

Marco looked up in surprise. "When you put it that way, they're actually four different machines with four different problems."
"Exactly!" Sandra said, now more animated. "And that paper jam in section C — it doesn't just happen randomly. It's always after product changeovers from light to dark packaging. Never the other way around."
Tom frowned. "Wait a minute. How do you know that for certain?"
"Because I have to run over to that thing every single time," Sandra replied with a wry smile. "After fifteen years, you recognize patterns, even if you can't prove them. It's as if the machine is allergic to that transition." She thought for a moment, then said, "I think this problem started in the spring of last year, but I'm not sure. It seems like it got worse around then."

Pedro felt the excitement growing. "This is gold, Sandra. We're going to turn your observations into a concrete problem definition that we can analyze. No more vague descriptions, but something we can really work with."

5.2. The Art of Specificity

Pedro grabbed a new marker and stood at the whiteboard. "Sandra, let's describe that paper jam in section C as precisely as possible. Not 'the belt is blocking,' but as specifically as possible. We need to define our 'Primary Effect' in a structured definition. That ensures we don't forget anything."

"What exactly is a primary effect?" Linda asked.
Pedro turned to her. "Good question. The primary effect is the problem you want to solve. In this case, we could say that the production loss is the problem. That's fairly clear, but the question is: why is that actually a problem? Suppose a process computer fails because water gets into it from a leak above it. Is the leak the problem, or is it the computer failing?

Depending on what you choose, different or more solutions are possible. In this case, it's pretty clear."
He drew a table on the board and began asking systematically:

5W+1H checklist:

What	What exactly is the problem?
Where	Where does it occur?
When	When does it happen?
Who	Who is affected?
Why	Why is it a problem?
How	How do you recognize it?

The team filled in the matrix together:
Primary Effect of failure on conveyor line three:

What	Paper jams
Where	Section C, between guide roller 6 and 7
When	2-3x per day, at product changeover light→dark
Who	Operators line three, production team
Impact	average 8 min downtime per incident at $200 per hour = $140,000/year loss due to downtime plus $100,000 in extra costs.
How Recognizable	Crumpling paper, complete blockage

Jake looked approvingly at the board. "Now we have something concrete to work with instead of 'that annoying belt.' This is the kind of specificity that management can do something with."
Linda was already working on her laptop. "And I can now collect specific data about this exact problem instead of drowning in generic 'conveyor belt failures'. By exporting the failure data from the EAM to spreadsheets, I can neatly categorize the different failure types."

Pedro felt that the team was beginning to grasp the power of a good problem definition. A failure only becomes solvable when you can describe it precisely. The difference between “the car isn’t working” and “the car won’t start at temperatures below zero degrees” is the difference between helplessness and a plan of action.

The Importance of What It Is NOT

"Now we're going to do something that might sound strange," Pedro said, "but it's crucial. We're going to systematically define what our problem is NOT. This Is/Is-Not analysis helps us rule out causes."
He drew two columns on the board:

What is the problem?	What is the problem NOT?
Section C conveyor line three	Other sections of the belt
Paper jams	Sensor failures or mechanical blockages
Since last spring	Before that
Product changeover light→dark	Product changeover dark→light
’s Morgens vroeg	Later in the day, less so
Dark packaging with coating	Light packaging without coating

Figure 5-1 Specific description of problem

Marco looked skeptical. "What do you mean? Why would you waste time on what it isn't?"
"Because," Pedro explained, "by systematically ruling out where the problem does NOT occur, we can eliminate causes. If it were mechanical, it would also happen at night. If it were general wear, it would occur with all paper types."
Sandra thought it over, methodically working through her experience. "You're right — never during night shifts. Never with light packaging only. Never during startup or shutdown. And never during dark-to-light transitions."
Tom began to see the implications. "That means we don't need to look for problems with the basic mechanics or general wear."

Linda added: "And it largely rules out electronic issues, because those wouldn't be time-dependent."
Marco nodded slowly. "So we're looking for something specific to the transition from light to dark paper. That narrows the search area enormously."

5.3. Detective Work: Gathering Information

"Now comes the detective work," Pedro said enthusiastically. "We're going to gather information from every possible source. I'll divide you into research pairs to systematically collect information."
He pointed to Sandra and Linda. "You two take the operational side. All the logbooks, production data, operator observations. Look for patterns that others might have missed."

He then turned to Marco and Tom. "You two dive into the technical documentation and maintenance history. Original drawings, modifications, all maintenance work from the last two years."
“Jake and I,” Pedro continued, “take the management information and external factors. Budget decisions, supplier changes, changes in procedures.”

Sandra still seemed a little uncertain. "What kind of information exactly? I mean, I see those jams every day, but how do I turn that into systematic information?"
Linda explained, "Everything that can help us understand why those jams occur. Operational logs, production data, environmental conditions. Think about temperature, humidity, time of day, who's on shift."

Marco was already busy making plans. "I'm going to dig up the original drawings of section C. And the maintenance history from the last two years. Every modification, every replacement."
Tom added, "I'll check all the electrical schematics. And those strange electrical stops Sandra mentioned – maybe they're connected to the paper jams."
"Don't forget," Pedro added, "we also need to verify whether those 'sharp contrasts' in our definition are actually true. For example: do the problems

really occur mainly or only during the transition from light to dark paper, or does that turn out to be incorrect when we look at the facts?"

Data Quality: The First Challenge

A week later the team met again. Linda looked frustrated as she opened her laptop.
“We have a challenge,” she said. “Our failure registration is a mess — the older data is inconsistent and difficult to interpret.”

"Fortunately," Linda continued, "the data from the last few months is much better structured thanks to the new registration system. I can definitely find patterns in that."
Marco slammed his fist on the table. “Can’t we just start keeping better records from now on?”

"The new system is already working well," Pedro said calmly. "But we also need to clean up the existing data and make it usable. We can't wait until we have perfect data - then we'll never get started."
He stood up and started writing on the board. "Linda, can you group the old data based on location and time? If different people report the same incident, we can merge those."

Sandra's Golden Observations

While Linda struggled with the historical data, Sandra shared something that only years of experience could provide.
"Something has caught my attention over the past few days," she said. "Those paper jams always follow the same pattern. First you hear a very soft squeaking sound — most people don't even notice it. Then, about five minutes later, the paper starts to crinkle at guide roller 7. And then — bam — complete jam."
Tom looked up from his charts. "A squeaking noise? From where?"
"I'm not exactly sure," Sandra admitted. "But it's always there. And it only happens with dark paper with that special coating. It's as if the machine is warning you before something goes wrong."

Marco's eyes lit up. "Wait a minute. Dark paper has different properties than light paper?"

Linda made a note. "I'm going to look into that. There may be specification differences we're not aware of. Coating, thickness, conductivity."
Sandra continued: "And it does seem worse in the morning. As if the machine is tired after the night."
Pedro felt the puzzle pieces slowly falling into place. Sandra's practical experience yielded clues that no data analysis could
give. The squealing noise, the timing, the correlation with paper type — these were valuable observations.

Marco's Finding

A few days later, Marco walked into the conference room, his laptop under his arm.
"Guys, I found something!" he called out. "Take a look at these maintenance records from two years ago."
All eyes turned to his laptop as he shared the screen. "Major overhaul, March last year. Guide roller 7 was replaced. But look here —" he pointed to a line in the report "— there's no clear documentation of all the steps that were performed."

Tom looked more closely at the screen. "What exactly do you mean?"
"Look," Marco said, barely able to contain his excitement, "it says here that the roller was replaced, but not what else was checked or adjusted. That could be quite relevant."
Sandra slapped her hand on the table. "March of last year! That could explain why the problem started back then!"
Linda was already typing. "I'm going to check that against my data. If the problem started then, it should be visible in the failure frequency."
Jake looked from one to the other. "You think something during that maintenance service caused these problems?"
"It could be," Pedro said carefully. "But we need to verify it before we draw conclusions. For now, it's a lead — a good lead, but not yet proof."

5.4. Puzzle Pieces Begin to Take Shape

Pedro walked over to the whiteboard where their findings were beginning to take shape.
"Let's stop for a moment and look at what we've gathered," he said, picking up the marker. "Sandra: squeaking sound before jams, only during light-to-dark transitions, worse in the mornings. Marco: possible change since last year's March maintenance. Linda: clear time patterns in the better data."

Jake leaned back in his chair. "It's starting to become a coherent story rather than a collection of isolated facts."
Tom was already busy making plans. "I'm going to look further into that maintenance session from March last year. Maybe there are more details to be found about exactly what was done."
Sandra nodded. "And I'm going to document those squeaking sounds systematically. With timestamp, paper type, ambient conditions."
Linda was busy working on her laptop. "I'm going to analyze all the data from after March of last year. If the problem started then, it should be visible in the failure frequency."

Systematic Observation Methods

"Before we continue," Pedro said, "I want to show you how we can use systematic methods to strengthen our information gathering."
He pulled a stack of forms from his folder and spread them across the table.
"Through structured observation, we can see patterns that would otherwise remain hidden," he explained. "Time, temperature, humidity, sounds — all systematically recorded."
Linda looked at the forms. "Can we use spreadsheets to find correlations between different factors?"
"Absolutely," Pedro replied. "Simple statistical analysis can reveal connections between various parameters. And trend graphs can make developments visible."

Tom studied one of the checklists. "If we can take specific measurements around guide roller 7 during the next maintenance visit, we can get more clarity."

Sandra looked relieved at the simple forms. "This I can actually understand. Just observe carefully and write down what you see."
"Exactly," Pedro assured her. "Your experience and observational skills are the foundation. These forms only help us systematically capture your insights and share them with the team."

Real-World Example from Experience

Pedro looked at the board full of information. "Let me give you an example of why this systematic approach is so important. I remember a case at a water treatment plant where pump station P-12 failed intermittently."

He drew a simple diagram on the board. "The maintenance team wanted to immediately overhaul the pumps - $30,000. But the maintenance engineer first conducted a thorough data analysis. She combined failure data, weather data, and maintenance schedules."
"Well?" Marco asked curiously.
"Pump failure correlated with heavy rainfall, but only at night. It turned out that automatic flush cycles had been programmed to run at night to save energy. During rainfall combined with backflushing, hydraulic pressure spikes occurred."
Sandra leaned forward. "And the solution?"
"Reprogramming the rinse cycles. Cost $500, reduced failures by 92%. That's why systematic analysis is so valuable — it prevents you from solving the wrong problems."

5.5. Verification and Validation of Information

"There is one more important aspect we need to pay attention to," Pedro continued. "Not all collected data is equally reliable. We need to verify our information with triangulation before we proceed."

He explained the method: "Triangulation means that you verify every important finding through three different sources. Only when all three indicate the same thing can you have confidence in it."
He pointed to the board where their findings were listed. "Look at our findings on timing and paper type:"

Triangulation example timing problem:

Source 1 (Observation)	Sandra - squeaking sound with dark paper, timing in the mornings
Source 2 (Technical Data)	Marco - maintenance service March last year
Source 3 (Failure Data)	Linda – increase in jams starting April last year, specifically at light→dark

Linda had a spreadsheet open. "I've done the timeline validation. The chronology checks out: maintenance service March last year, increase in jams starting April last year, specifically during light-to-dark transitions."
Tom added, "And the technical plausibility: a change during maintenance can indeed cause new problems."
Marco was busy sifting through paper maintenance reports. "I've gone through all the maintenance reports from that period. There are indeed only limited details about what was actually done."
Pedro emphasized: "This triple verification tells us that our findings are solidly grounded. Without triangulation, it would remain guesswork."

The Next Step: Root Cause Analysis

"Next week," Pedro announced, "we're going to convert this information into hypotheses that we can test systematically. We now have a sharp problem definition and a wealth of information. Now we're moving on to the Root Cause Analysis itself."
Sandra seemed much more confident than at the beginning of the process. "I have faith in it. For the first time, it feels like we're working systematically. My experience counts, but it's also supported by data and methodology."

Linda closed her laptop. "And I'm starting to understand how qualitative observations and quantitative data can reinforce each other."

Pedro looked with satisfaction at his team. They had learned what every reliability engineer must master: the art of converting vague complaints into specific, analyzable problems.
"Wednesday," he concluded, "you'll get to know the different methodologies for Root Cause Analysis. We go from clues to evidence, from suspicion to certainty. Because what we have now are good leads, but not yet definitive proof."

The team was ready for the next phase. They had learned the fundamental lesson that problems only become solvable when you can describe them precisely and analyze them systematically.
The journey from reactive firefighting to proactive structural improvement had begun. And it started with the simple but powerful insight that you cannot solve what you cannot define.

6. Methodologies for Root Cause Analysis

6.1. Systematic Approach to Root Cause Analysis

Now that you have learned how to define problems clearly and collect information systematically, you are ready for the core of structural improvement. If you want to resolve a failure or other problem permanently, you must work systematically and carefully. Where possible, you must eliminate the root cause of the failure. To do that, you obviously need to know the cause first. Often, people try to solve a failure without having thoroughly investigated what the cause is. They draw conclusions too quickly. They then waste time and money by trying out a solution that turns out not to work. This is not only a waste of time and money, but it can also cause new problems.

An important principle is "if it ain't broken, don't fix it." This means that things that are working well should be left alone as much as possible. Every time you work on them, there is a chance that they will no longer function properly afterward. If you make changes to an asset, the risk is even greater. There is then a chance that it will no longer work at all, or become dangerous, while it is not easy to revert to the old situation. It is therefore important to think things through carefully before implementing a solution.

From Information to Methodology

It was Wednesday morning. Pedro stood in front of the whiteboard in the conference room with all the information the team had gathered over the past week. Stacks of documents, photos of the conveyor belt, maintenance reports, and handwritten notes were neatly sorted on the table.
"Okay team," Pedro began, pointing to the assembled information, "we now have mountains of information about our paper jam problem. Sandra's observations from the shop floor, Marco's maintenance history, Linda's data analysis, Tom's electrical diagrams, Jake's production data. "But gathering information is one thing, systematically analyzing it is another."

Jake Jensen, the production supervisor, looked up from his notes. "You "means that we're now going to figure out what information is actually important?"
Sandra the experienced operator of line three, sighed. "I hope we're not going to brainstorm at a whiteboard again and just shout out whatever we think the cause is. We've done that so many times already."

Pedro smiled. "That is exactly what we are NOT going to do. We're going to work methodically. There are proven techniques to get from symptoms to root causes. And I'm going to teach you how to systematically identify all possible causes without missing a single one."
Marco, the mechanical technician, raised an eyebrow. "Let me guess - more spreadsheets?"
"Among other things, yes," Pedro replied. "But smart spreadsheets that help us work systematically rather than just throwing things out there."

6.2. Simple Methods for Root Cause Analysis

For most organizations, it is wise to start with proven, accessible methodologies before moving on to more complex approaches. The search for the root cause is done using a methodology called Root Cause Analysis (RCA). There are various approaches to RCA. Basic RCA methodologies are suitable when you expect fewer than three possible root causes, are analyzing simple systems, or are investigating mechanical problems.

6.3. 5-Why Method: Step-by-Step Explanation

"We'll start with the 5-Why method," Pedro explained, picking up a marker. "Simple principle: keep asking 'why?' until you reach the root cause."
Tom, the electrical technician, chuckled. "Sounds like a four-year-old toddler."
"Exactly!" Pedro exclaimed. "Toddlers are natural root cause analysts. Adults stop asking questions too soon. Just look at this example."
He began writing on the whiteboard:

Problem: Car does not start
"Step 1: Define the problem clearly. 'Car won't start' is still somewhat general. What does it do and what doesn't it do? For example, do you hear the starter motor turning? The better you understand how something works, the more specifically you can define this description can make. Let's assume the starter motor does not turn.

Step 2: Ask the first 'Why?' Why doesn't the car start? What do you all think?"
"Dead battery?" Linda Wang, the quality engineer, suggested. "Good.

Step 3: Keep asking why. Why is the battery dead?"
"Lights stayed on," Sandra said.
"Why did the lights stay on?" Pedro continued.
"Forgot to turn it off," Marco added.
"Why forgotten?"
"No routine?" Jake tried.
"Why no routine?"
"No checklist when getting out of the car," Tom concluded.
Pedro put down the marker. "Exactly! From 'car won't start' to 'no checklist'. That's a completely different solution than 'buy a new battery'."
Linda nodded enthusiastically. "I see the difference. If you buy a new battery, you're solving the symptom. If you make a checklist, you're addressing the underlying cause."
Sandra leaned back in her chair. "I recognize that. With jams, we always think first about fixing the machine, not about why it happens in the first place."

"Step 4: Don't stop until you reach an actionable cause," Pedro continued. "Even then, it's often useful to keep asking further questions."
Marco raised his hand. "But how do you know when to stop asking why? You can go on endlessly."
"Good question! You stop when further 'whys' take you outside your sphere of influence, or when you reach a cause that you can practically address. In the car example: creating a checklist is something you can do. 'Why do people forget things?' takes you into psychology and neurology — that goes too far for this problem."

Applying 5-Why to a Conveyor Belt

"Let's try applying it to our problem," Pedro suggested. "Why do we have paper jams in section C of conveyor line three?"
Sandra thought for a moment. "Because the paper jams between guide roller 6 and 7."
"Why does it seize up?" Pedro asked.
"Too much friction?" Marco suggested.
Tom shook his head. "Not always. Only with dark coated paper. With light paper it rarely happens."
"Interesting," Pedro said. "Why does coated dark paper have more friction?"
Linda looked at her notes. "We don't actually know for certain. We suspect static electricity from the coating, but..."
"Stop!" Pedro interrupted her. "This is where you see the problem. We're going from facts to assumptions. The 5-Why method only works if every answer is based on evidence, not assumptions."

Sandra frowned. "So we have to prove first that it's static electricity?"
"Exactly. That's why we need those verification steps that we'll work out later. Guessing at causes leads to solutions that don't work."
Jake made a note. "So we're pausing the 5-Why until we have more evidence?"
"That's not necessary," Pedro said. "We use it to gather hypotheses and verify each step later."

Practical Application of the 5M Model

Pedro turned to a new section of the whiteboard. "The 5-Why method is powerful, but can lead to tunnel vision. That's why we also use the 5M model to make sure we're considering all possible causes."
"5M?" Jake asked. "Man, Machine, Medium, Material, Method. Every possible cause falls under one of these categories."
Pedro wrote the five words in columns on the board. Linda's eyes lit up. "And that prevents us from only looking at technical causes?"

"Exactly! Research shows that approximately 50% of all failures are caused by human factors, but many organizations focus only on technical causes."

He pointed to each column. "Now we're going to systematically go through each category to make sure we don't miss anything. But no brainstorming — we're working through a checklist. Sandra, you take the Human category. Use this list."

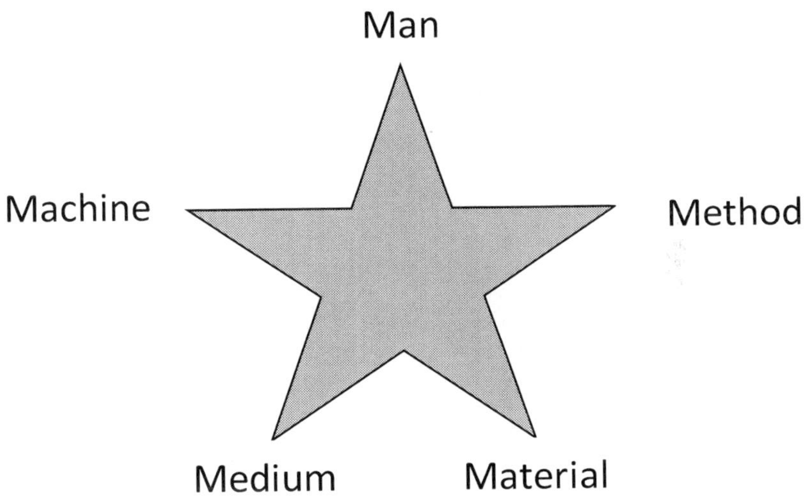

Figure 6-1 The five M's

People-category systematic checklist:

- Competency: Have all operators been trained for this specific process?
- Procedures: Are procedures current and available?
- Communication: Is information being passed correctly between shifts?
- Work pressure: Are there time-pressure situations that cause errors?
- Physical condition: Fatigue, concentration, physical strain?
- Supervision: Is adequate guidance available?

Sandra went through the list methodically. "Competency... there's a gap there. We do have training for the machine, but not specifically for product changeovers from light to dark paper."

"Procedures... they exist, but they're from last year. We have a new coating supplier."

"Communication... Jake, is there follow-through on problems during shift changes?"

Jake shook his head. "Not systematically. Usually only after there's just been a jam."

"Perfect," Pedro said. "Now Marco, you work out the Machine category."

Machine Category Systematic Checklist:

- Mechanical condition: Wear, alignment, vibrations of components
- Calibration: Are the settings correct and recently verified?
- Maintenance History: Which repairs have been carried out recently?
- Design limits: Is the machine operating within specifications?
- Sensors: Are all detection systems functioning correctly?
- Lubrication: Was maintenance performed according to schedule?

Marco worked through the list systematically. "Mechanical condition... I noted vibrations in roller 6. And I adjusted that alignment last month after that failure."

"Calibration... when were the roll pressures last checked? Linda, is that in your maintenance reports?" Linda flipped through her papers. "Six months ago. And back then only for standard paper, not for coated variants."

"And the installation is just getting a bit older too," Marco said. "Makes sense that things don't work as well anymore."

"Watch out!" Pedro replied immediately, "Time is not a cause. Wear, corrosion, and the like are causes. These do indeed occur over time, but time itself is not a cause. If you write that down in your analysis, you can't do anything about it either. Against corrosion, wear, and other causes that are time-dependent, you usually can, but you need to name them specifically."

> Time is not a cause of failures. Causes develop over time, but not as a result of time.

Tom took the next category. "Medium, this is another word for Environment... temperature and humidity vary considerably. In the afternoon it gets warmer from the sun on the roof. And that new ventilation has changed the airflow."

Linda picked up the Material file. "The coating supplier did indeed change. Different chemical composition, different thickness. And dark pigments have different properties than light ones."

Jake wrapped up with Method. "We have no specific procedure for product changeovers. Operators determine for themselves when they adjust settings. And inspection intervals haven't been adapted to the new material."

5M ROOT CAUSE INVENTORY - PAPER JAMS SECTION C	
MAN	• Operator forgets to adjust settings • Insufficient Training on Product Changeovers • No communication between shifts
MACHINE	• Vibration of guide roller 6 due to wear • Guide roller 7 not grounded • Roll pressure not calibrated for 6 months
MEDIUM	• Low humidity early in the morning • Temperature variation due to sun on roof • New ventilation changed airflow

MATERIAL	• Coating new supplier (6 months ago) • Dark pigments, different properties • Paper thickness varies within batch
METHOD	• No specific procedure changes • Control intervals not adjusted • Operators determine timing adjustments themselves

Pedro stepped back and looked at the whiteboard. "In fifteen minutes you've identified fifteen concrete, verifiable causes. Without the 5M model, we might have only looked at the machine."

Sandra shook her head. "I have to admit, normally I really do only think about the mechanical causes. But those procedures... that can genuinely be a factor."

Tom studied the complete table. "This is a good inventory, but it doesn't yet show how things relate to each other."

"That's right," Pedro said. "That's why we're now moving from this flat list to a deeper analysis."

6.4. From Flat List to Causal Chain — the Fault Tree

Two ways to display the same analysis

Pedro opened a new document on his laptop. "We now have fifteen possible causes on a list. But a list doesn't tell you what the cause of a cause is."

Marco frowned. "What do you mean?"

"Look at 'guide roller 7 not grounded' on our list. That's listed there as a fact. But why is that roller not grounded? And why wasn't that noticed? And why is there no system in place to prevent that?"

Sandra began to see it. "You want to dig deeper."

"Exactly. And for that we use a Fault Tree. It's essentially the same analysis as our 5M list, but in a format that shows causal relationships."

Linda looked up with interest. "What does that look like?"
Pedro projected a new table on the screen. "Instead of fifteen separate items, we now arrange them in columns from left to right. Each column is one level deeper in the 'why' question."

The Logic of the Fault Tree

He began to explain, pointing to the columns with his finger:

"Level 1 - The primary problem: Paper jams in section C. That's where it all begins.
Level 2 - The Direct Mechanism: What happens physically? Too much friction between the paper and the roller.
Level 3 - The physical cause: Why is there too much friction? Static electricity attracts the paper.
Level 4 and beyond - Going deeper and deeper: Why is there static electricity? Because the roller is not grounded, the humidity is low, the coating is conductive."

Tom looked at the diagram. "So you read from left to right, from symptom to cause?"
"Exactly. And when you see multiple rows beneath each other in the same column, means parallel causes - they all contribute to the same problem."
Jake saw the value. "This makes it immediately clear which causes are related and which stand independently from each other."
Marco nodded. "Our 5M list was the ingredients. This is the recipe - it shows how everything works together to cause the problem."

Events and Circumstances

"An important insight," Pedro said, "was for me once that you shouldn't only look at events, but also at circumstances."
"What do you mean?" Marco asked.
"People often tend to look at what went wrong at the moment the problem occurred. If there has been a fire somewhere, we say it was because someone threw away a cigarette."
"Can't you do that too?" Sandra asked, not understanding.
"That is indeed an event that can lead to a fire, but only under the circumstance that flammable material is present." Pedro clarified. "An event is

something that occurs at a specific moment. A circumstance has already existed for some time. You can therefore see it and counter it all along. It is perhaps easier to ensure that no flammable material is present somewhere than to ensure that nobody throws away a cigarette. If your emergency stops don't work, that is a circumstance; if your pipes are rusty, that is a circumstance. You can prevent a great many problems by making sure the circumstances are in order."

Identifying Microsteps — The Intermediate Links

Tom studied the Fault Tree carefully. "I see that we go from level 1 to level 4, but between level 2 and 3, for example, there is often more than one step happening. Are we keeping track of that as well?"
"Excellent question," Pedro replied. "Often there are intermediate steps between a cause and the ultimate problem. If you miss those, you may miss other possible solutions."
He drew an example on the whiteboard:

1. Without microsteps: Operator forgets setting → Jam occurs
2. With microsteps: Operator forgets setting → Roller pressure incorrect → Paper tension changes → Static charge buildup → Paper sticks to roller → Jam occurs

"Do you see the difference? If you only know the first and last step, the only solution seems to be better training. But if you know the microsteps, you see more points of intervention."
Sandra's eyes lit up. "You could install a sensor that measures the tension. Or apply an antistatic treatment. Or adjust the rollers differently."

"Exactly! Every micro-step is a potential place for a solution. In the Fault Tree, we try to make these intermediate steps visible by adding extra levels where needed."
Jake took notes. "So for every causal chain in our Fault Tree, we need to think about the complete sequence of events and circumstances?"

"That's right. And often you then discover that what you thought was one direct cause is actually multiple steps occurring one after another. Each step is an opportunity to interrupt the problem."
Linda nodded in understanding. "So if in our Fault Tree we go from 'static electricity' to 'paper sticks,' we need to ask ourselves: what exactly happens between those two moments?"
"Exactly. Sometimes those microsteps are so fast or subtle that you overlook them. But they are crucial for finding innovative solutions."

When do you use which view?

Sandra asked the practical question: "When do you use the 5M list and when do you use the Fault Tree?"
Pedro thought for a moment. "Good question. They serve different purposes:

Use the 5M list when you:

Brainstorms with the team
Quickly collect all possible causes
Make sure you don't forget any categories

Use the Fault Tree when you:

Wants to understand causal relationships
Do you want to see how deep a problem goes
Must communicate to management
Priorities want to set based on impact"

Linda added: "So basically: 5M for collecting, Fault Tree for analyzing?"
"Exactly. They complement each other. The 5M ensures you cast a wide net, while the Fault Tree helps you dig deep."

The Conveyor Belt Fault Tree - Level 1 to 3

"Let's convert our 5M findings," Pedro suggested. He began filling in the table, with input from the team.
It took about an hour before they had an overview of all the main branches down to level 3.

Sandra looked surprised at the overview. "I hadn't thought it could be this structured. Normally I just say 'the conveyor's jammed again' and then we just fix whatever."
Jake saw the strategic value. "With this overview, I can go to management and say: these are our main problems, this is how they're connected, and this is what we're going to address."
Marco counted the parallel causes. "Fourteen different factors at level three that all contribute. If we were to address all of them..."
"That's exactly why we prioritize," Pedro interrupted him. "But before we do that, I want to show you how deep a Fault Tree can go. Because level three is not the end of the story."

6.5. In-Depth Analysis – How Deep Do You Need to Dig?

From Technical Cause to Organizational Root Cause

Pedro looked around at the team. "Our Fault Tree stops at level three right now. Those are all technical, physical, or procedural causes. But I want to show you what happens when you keep asking questions."
He opened a new table. "Let's go all the way down one branch — the grounding branch. Because there's more to it than 'technician forgot to secure the wire.'"
Sandra leaned forward. "How much deeper can you go?"
"That depends on the problem. But for this problem - certainly three or four levels deeper."

> The term 'Root Cause' does not mean that a problem has a single cause. This is a persistent misconception. On the contrary, it means that every problem has an entire 'root system' of causes. If you turn the analysis sheets on their side, you can see how a wide range of 'small' causes form increasingly thicker roots until you are finally left with just one consequence at the top.

PRIMARY EFFECT	LEVEL 1	LEVEL 2	LEVEL 3
Paper jams section C **Frequency:** 78x per month **Impact:** $264,000 per year	Excessive friction between paper and guide roller 7	Static electricity (2.3 kV measured) attracts paper to roller	Guide roller 7 not grounded - charge cannot dissipate
			Low relative humidity in the morning (<35% RH)
			HVAC system has insufficient humidification capacity
			Coating of dark paper is more electrically conductive than light paper
			When switching light→dark, sudden discharge occurs via conductive coating
			New supplier (6 months ago) uses different coating composition
	Operator does not adjust settings during product changeover	Procedure for light→dark paper changeover unknown to operator	Procedures are outdated (last year), not updated for new coating
			Procedure change not communicated to all shifts
			Insufficient training for product changeovers with coated material
			No communication between shifts about abnormal conditions
	No early warning - problem escalates unnoticed	No temperature monitoring of guide rollers	When line was purchased (2019), no condition monitoring was specified
			Business case focused on purchase price, not on lifecycle costs
		No vibration monitoring	See condition monitoring above - same root cause
			Monitoring was seen as "nice to have" not as essential
How to read: Left to right = from symptom to cause. *Multiple cells stacked within one column = parallel causes that contribute together.*			

Figure 6-2 Fault tree analysis levels 1 through 3

Developing the complete grounding chain

Pedro began filling in the table, with the team thinking along:

"Level 4: Guide roller 7 not grounded. We know that.

Level 5: Why wasn't it grounded? The technician didn't reinstall the ground connection during the overhaul in March of last year. But also: work planner James didn't check it."

James, the work planner who had joined them, nodded guiltily. "That's right. I only check whether the mechanical work is complete, not whether the electrical aspects are correct."

"So we have two parallel causes at level five," Pedro continued. "And now the question: why did neither the technician nor James check it?"

The team thought it over. "At our place, the mechanical technician doesn't have a checklist with electrical items. And vice versa," Marco said.

"Level 6:" Pedro wrote, "No integrated checklist that combines mechanical and electrical checks."

Tom added: "And the technician probably thought: electrical will handle that. No ownership over the total asset."

"Also level six," Pedro confirmed. "And why don't we have integrated checklists?"

Linda knew the answer: "Because mechanical and electrical work separately in our EAM system. You create either a mechanical work order or an electrical work order, but not a combined one."

"Level 7:" Pedro wrote, "Mechanical and electrical maintenance separated in EAM system."

Jake looked thoughtful. "And that's because when we implemented the EAM system in 2018, we simply digitized our old paper systems. Everyone wanted to go live quickly; nobody wanted to redesign the processes."

"Level 8:" Pedro wrote, "During EAM implementation, speed was chosen over process integration."

Sandra shook her head. "And why did everyone want to move fast? Because every department wants to do its own thing. Nobody wants to cooperate with other departments."

"Level 9:" Pedro wrote slowly: "Silo mentality: disciplines operate as separate islands without cross-functional collaboration."
The team fell silent. They looked at the complete chain on the screen.

Level 4	Level 5	Level 6	Level 7	Level 8	Level 9
Guide roller 7 not grounded. Charge cannot dissipate	Technician did not reinstall grounding during overhaul March 2022	No integrated checklist combining mechanical and electrical checks	Mechanical and electrical maintenance separated in EAM system	During EAM implementation, speed was chosen over process integration	Silo mentality: disciplines operate as separate islands without cross-functional collaboration
		Technician assumed 'electrical will handle that'. No ownership of the complete installation			
		No training in cross-functional thinking			
	Work planner did not check work order for electrical aspects	He only checks completeness within his own discipline	Work planner job description contains no cross-check responsibility		
			No validation step in workflow where both disciplines sign off		

Figure 6-3 In depth fault tree analysis of grounding

The Progression Through the Levels

Marco was the first to speak. "We started at 'wire isn't connected' and ended up at 'nobody works together'. That's quite a leap."
"Do you see the pattern?" Pedro asked. "Level 4 is technical – a physical component is missing.
Level 5 is execution — people did not complete their work.
Level 6 is procedural — systems to prevent this are missing.
Level 7-8 are organizational – the structure and decisions that lead to failed procedures.
Level 9 is cultural - the underlying mindset."

Linda typed along. "So the deeper you go, the more you shift from 'what went wrong' to 'why does it keep going wrong structurally.'"
"Exactly. But be careful," Pedro warned, "this is not a fixed formula. Not every problem follows neatly physical→technical→procedural→organization→culture. Some problems skip levels, others branch differently. The

important thing is: keep asking questions until you reach causes that you can and want to address."
Tom looked at the parallel branches in the table. "I see that some level 5 causes jumping directly to level 7, skipping level 6."
"Good point," Pedro said. "Take James's job description, for example — that's an organizational document, not a procedural problem. Reality is messy, not linear."

The Exponential Impact of Deep Solutions

Jake asked the crucial question: "But if you solve level 4 versus level 9 - what's the difference?"
Pedro smiled. "That's exactly the point. Let's work through the numbers:

If we address level 4 - restoring the grounding:
Investment: $2,000
Impact: Resolves this specific problem
Problems solved: 1
Risk of recurrence: High – it will happen again at the next overhaul

If we tackle level 6 — creating integrated checklists:
Investment: $10,000
Impact: Does this problem prevent all future overhauls
Problems resolved: 10-15 per year
Risk of recurrence: Low – the system prevents it

If we address level 9 — breaking down silo mentality:
Investment: $250,000 over three years
Impact: Eliminates the underlying cause of dozens of different problems
Problems solved: 100+ per year
Risk of recurrence: Very low – the culture is changing"

Sandra let out a soft whistle. "So the deeper you solve it, the more problems you prevent?"
"Not linear — exponential," Pedro emphasized. "A technical solution is additive: you solve one problem. A system solution is multiplicative: you

solve many problems. A cultural solution is exponential: you eliminate entire categories of problems."

"But level 9 also costs a lot more time and money," Marco said.

"That's why," Pedro said, "you work in parallel on multiple levels. Level 4 for quick wins - we'll restore that grounding next week. Level 6 for the medium term - we'll develop integrated checklists over the coming months. Level 9 for structural transformation - that's a multi-year trajectory that management needs to be involved in."

When do you stop digging deeper?

Tom had a practical question. "How do you know when you've gone deep enough? You can keep asking 'why' endlessly."

Pedro wrote four criteria on the whiteboard:

Stop digging deeper when:

- Actionable level reached – You arrive at a cause that you can actually address within your mandate and resources
- ROI Optimum - Going deeper costs more time and money than it yields
- Outside your scope - You encounter organization-wide systems that go beyond this project (but do report them to management)
- Repeating pattern - You see the same root cause recurring from different analysis paths

Linda nodded in understanding. "So in our case: level 9 is clearly outside our scope as a team. But we can flag it and handle levels 4 and 6 ourselves."

"Exactly," Pedro confirmed. "And that is also what we are going to do. We report to management: 'We are solving this problem technically, we are preventing recurrence with procedural improvements, and we see a pattern of silo thinking that you need to address organization-wide'."

Jake made a note. "That's a stronger message than 'we tightened a wire.'"

Barrier Thinking and Change Analysis

Jake brought a different perspective. "But even if we know what the direct cause is — like that missing ground connection — we also need to ask: what should have prevented this?"

"Barrier thinking," Pedro confirmed. "For every cause you ask: which line of defense should have stopped this? And why did it fail?"

He pointed to level 5 in their Fault Tree. "The technician forgot the grounding.

- Barrier 1: the technician himself should have checked it - failed.
- Barrier 2: the checklist should enforce it - does not exist.
- Barrier 3: the work planner should validate it — only performs a mechanical check.
- Barrier 4: the EAM system should have issued a warning - is not integrated."

Sandra saw the pattern. "Four lines of defense, all failed."

"And that's no coincidence," Pedro said. "They failed for the same underlying reason: silo thinking."

Tom added another insight: "And Change analysis helps too. Problems often arise because something changed. Our new coating, the new ventilation, the supplier switch - all changes that introduced new problems."

"Exactly," Pedro confirmed. "If something always worked well and now it doesn't, ask yourself: what has changed? That often leads quickly to the trigger."

They continued filling in their spreadsheet. Figure 6.4 shows a summary of the remaining columns of this spreadsheet.

Possible cause	Barrier failed	Recent change	obabil	Status	Resp.	Deadline
Operator forgets settings after product changeover	No checklist procedure	-	7	To be investigated	Sandra	March 15
Vibration roller 6 due to bearing wear	No vibration monitoring	Roller tension adjusted last month	8	To be investigated	Marco	March 12
Static charge due to low humidity (< 35% RH)	No antistatic measures	New ventilation last year	6	To be investigated	Tom	March 18
Guide roller 7 not grounded since overhaul March 2022	No electrical checklist for mechanical work	Overhaul March 2022 - grounding not restored	9	CONFIRMED Evidence level 3	Tom	Done
New supplier coating has different properties	No supplier validation	Supplier change 6 months ago	9	CONFIRMED Evidence level 2	Linda	March 10
No procedure for product changeovers	Procedure was available but unclear	Procedure change not implemented	8	CONFIRMED Evidence level 3	Jake	Done

Figure 6-4 Barrier and change

6.6. Avoiding Premature Conclusions

The Pitfall of Quick Judgments

Marco suddenly went quiet. "I have to confess something. Last year we had a similar problem on line 7. I was so sure it was a bearing problem - all the symptoms matched perfectly."

Tom looked up curiously. "And?"

"Management gave me budget for new bearings. We did a complete overhaul, really went through everything thoroughly." Marco shook his head. "After installation: not a single improvement. It turned out to be a software problem in the control system. The vibrations I interpreted as bearing wear were caused by incorrect timing in the drive control."

Sandra recognized the story. "That happens to me too. When you have years of experience, you think you can see the cause directly. You see patterns that may not be there."

"That's exactly why we work systematically," Pedro emphasized. "Experience is valuable, but it can also blind you to other possibilities."

Psychological Pitfalls

Linda had made a list. "From the literature, I know of various cognitive biases that we make:

- Confirmation bias: We primarily seek out information that confirms our initial ideas.
- Availability bias: We consider solutions based on recent experiences.
- Anchoring effect: We place too much value on first impressions.
- Overconfidence: We overestimate our ability to make the correct diagnosis."

Tom grinned. "Like my tendency to always look at electrical problems first, because that's my area of expertise."

"Our brain does exactly what it's supposed to do — quickly recognize patterns and make decisions efficiently," Pedro said reassuringly. "But with complex technical problems, those snap judgments can be misleading."

Strategies Against Premature Conclusions

Sandra looked at her notes. "How do you prevent yourself from falling into those traps?"

Pedro gave concrete strategies:

1. "First: be aware of your own thought processes. Acknowledge that you have bias. We all have preferences and experiences that influence us.
2. Second: always generate multiple possible explanations before choosing. Our 5M method and Fault Tree force you to do this.
3. Third: actively seek out counter-evidence. Don't just ask "does this add up?" but also "what could disprove this?"
4. Fourth: work in diverse teams. Different perspectives help expose each other's blind spots.
5. "Fifth: document your reasoning process. When you look back later, you can see where you made errors in your thinking."

Marco looked at the Fault Tree on the screen. "That's why all those columns and that systematic approach. They force you to think slowly and methodically."

"And be extra critical of 'favorite solutions'," Pedro added. "Things you've seen work before. They're tempting, but may not apply to this specific problem."

Jake summed up: "So systematic methodology protects us from our own biases and quick conclusions?"

"Exactly. It forces us to slow down, look broadly, and verify before we invest."

6.7. Team Composition for Effective Analysis

The Power of Diversity

Pedro looked around the table at his team of six. "We're not here together by chance. A good RCA requires different roles and perspectives."

He pointed to himself. "I facilitate – I monitor the methodology, make sure we stay systematic, and ensure we don't jump to conclusions too quickly."

Linda nodded. "I'll take care of the data analysis. Looking for patterns in failure data, calculating correlations."

Sandra raised her hand. "I bring the practical knowledge. What really happens on the shop floor, what signals do operators see before a jam occurs."

Marco rolled his shoulders. "Mechanical expertise. If there are technical causes, I need to be able to recognize them and assess whether proposed solutions are realistic."

Tom tapped on the table. "Electrical systems and automation. Especially that grounding theory and all the sensor-related issues."

Jake made one final note. "Management perspective. Making sure our solutions are realistic and implementable within budget and schedule."

"Perfect!" Pedro said enthusiastically. "Diverse expertise, different perspectives, different experience levels. That's exactly what you need for a good RCA."

Core Roles in an RCA Team

Pedro explained which roles are essential:

"The facilitator guides the process. He does not need to be a subject matter expert, but monitors the methodology, ensures that all team members can make their contributions, keeps the focus on the process, prevents premature conclusions, and documents the progress."

Linda asked, "So the facilitator can come from outside the department?"

"Absolutely. Sometimes that's even better, because you don't have any preconceived ideas about the problem."

Technical experts bring in-depth knowledge of the systems involved. They can quickly assess whether a solution is technically feasible and help avoid pitfalls that are only visible to someone with years of experience.

Operators who work with the equipment daily offer a practical perspective from the shop floor.

Maintenance personnel such as breakdown mechanics and technicians provide insights from the maintenance perspective.

"Sometimes you also need external specialists — vendors or consultants who have specific expertise."

Criteria for Team Composition

Jake asked, "How do you determine who you invite?"

"Diversity of expertise is crucial," Pedro replied. "You want a mix of technical, operational, and process knowledge. Different levels — both frontline personnel and management — guarantee different perspectives."

Linda added: "And independence. Some team members who were not directly involved in the incident can look at it more objectively."

"Availability must be guaranteed," Pedro continued. "Team members must have sufficient time.

"Commitment is crucial - choose employees who are motivated to contribute."

Sandra looked around the table. "And the team size?"

"Keep it manageable, usually 5–8 people. If the group is larger, some people will no longer actively participate, or discussions will become far too lengthy. For larger problems, subteams can be formed that merge later."

Tom saw the practical value. "And for other problems, we adjust the composition?"

"Absolutely. For an electrical problem you need different experts than for a process problem. The core roles remain — facilitator, operators, maintenance — but the specific experts rotate."

From Methodology to Action

Pedro wrapped up the meeting: "We now have the methodologies to systematically analyze:

- The 5M model ensures that we collect broadly and don't overlook any category
- The Fault Tree lets us dig deep and see causal chains
- Verification principles keep us honest — we don't guess, we prove
- Diverse teams protect us against blind spots and biases

Marco nodded. "And now?"
"Now we're going to apply these methodologies for real," Pedro replied.

In chapter 7, we follow the team as they systematically verify their hypotheses. From suspicions to proven facts. From theory to evidence.
Sandra was ready for action. "Finally moving from talking to doing."
Linda closed her laptop. "With the certainty that we're working methodically instead of randomly trying things."
Tom stood up. "Let's start by verifying that grounding. I want to make sure our hypothesis is correct before we invest."
Pedro smiled. "Exactly the right attitude. Next week we start the verification process. From analysis to evidence, from evidence to solutions."

7. Proving Possible Causes

7.1. From Theory to Evidence

It was Monday morning, week four of the RCA project. Pedro sat with his team around the table in the conference room at InduPack Solutions. In front of them lay a printed spreadsheet with fifteen possible causes according to the 5M model. The theoretical work was done, but now the real work was beginning.

"Okay team, we now have a nice list of suspects," Pedro began, holding up the list. "Sandra thinks it's humidity, Marco points to that missing ground connection, Tom is talking about static electricity, Linda sees patterns in the data."
Sandra shifted uncomfortably in her chair. "I do have some ideas about what's going wrong."
Jake tapped his pen impatiently on the table. "We all have theories. But management doesn't want theories - they want results."
"That's exactly the problem!" Pedro said calmly. "We're losing $240,000 a year. If we address the wrong cause, we're wasting time and money. So now we're going from guessing to knowing. From theories to proven facts."

Marco looked skeptical. "How? We can hardly shut down the entire production line for experiments."
"We're becoming detectives," Pedro replied with a smile. "Every hypothesis we will systematically test. Some through observation, others through small experiments, still others through data analysis. But by the end of this week we won't have suspicions anymore — we'll have evidence."

The Verification Process

Now that you have identified the various possible causes, it is time to determine which of them are actually responsible for the problem. This verification process distinguishes guessing from knowing — and that

makes the difference between a solution that works and one that does not.
The verification process at InduPack began systematically. Pedro divided the tasks across the team, so that each hypothesis would be examined from different angles. Sandra would convert her years of experience into systematic observations. Marco and Tom would conduct physical tests. Linda would immerse herself in the historical data. Jake would provide practical support.

"Everyone gets an observation form," Pedro said, handing out the papers. "We're going to document everything: time of day, paper type before and after changeover, weather conditions, number of people on shift, preceding sounds or vibrations, and the exact moment of the jam."

Sandra's Methodical Observations

Sandra was given the assignment to convert her years of experience into systematic observations. She was used to working "by feel," but now she had to become objective.
Linda looked at her encouragingly. "Sandra, you have the best feel for the machine. But we need to convert your experience into objective observations."

"What do you mean?" Sandra asked. "Instead of 'I feel like something is going wrong,' we're going to document: exactly when, under what circumstances, with what preceding signals," Pedro explained.
Three days later, Sandra came into the conference room excitedly. "Pedro, I suspected it before, but now I know for sure! Every jam starts with a specific beeping sound — very soft — exactly five minutes before the real problems begin." Tom looked up questioningly. "Have you been able to determine where exactly that sound is coming from?"
"I'm not sure exactly, but it's always there," Sandra replied enthusiastically. "And it only happens during the transition from light to dark packaging. Never the other way around. And never within the same paper type." Marco clapped his hands. "Sandra, that's gold! You've found evidence of an early warning signal."

Marco's Systematic Experiments

Marco wanted to test his grounding theory, but scientifically this time. He had always worked on intuition and experience, but now he needed proof. "If Tom is right about static electricity from poor grounding, then I should be able to measure that," Marco said during the next team meeting.
Pedro nodded approvingly. "How are you going to approach that?"
"I'm temporarily installing extra measurement equipment. Then we'll test under various conditions: high and low humidity, different paper types, different times of day." Tom was curious. "And if you measure voltage on guide roller 7?"
"Then we have our proof. If I don't measure any voltage, then the grounding theory is wrong."

Marco set up a systematic test arrangement. He connected a multimeter to guide roller 7 and took measurements every ten minutes throughout a complete shift. In parallel, Sandra observed sounds and jams, while Linda recorded all production changeovers.
After the first test day, Marco came to the team excitedly. "Guys, look at this! With low humidity and coated dark paper, the voltage rises to 2.3 kilovolts!"
Tom let out an admiring whistle. "That's enough for quite a shock."
"And look here on my oscilloscope — that voltage discharges every 45 seconds. That explains the vibrations we're measuring!" Sandra looked triumphant. "And the squealing noise! That must be static discharge."

Linda’s Data Detective Work

Linda dove into the historical data in the meantime to prove the patterns. She had exported all the jam data from the last two years out of the EAM system and was now looking for cross-correlations.
"I've cross-referenced all the jam data from the last two years with weather data from the local weather station," Linda reported after days of spreadsheet work. Jake looked up curiously. "And?"
"There's a strong correlation with wind direction. East wind: 340% more jams. West wind: normal frequency." Sandra frowned. "Wind direction? How can wind affect our conveyor belt?"

"Not the wind itself. An east wind brings cold, dry air. A west wind brings moist air from the sea." Tom was beginning to understand. "So indirectly, the weather affects our indoor humidity?"
"Exactly! And look here — seasonal patterns. Winter has 60% more jams than summer, despite identical production. That confirms the humidity hypothesis." Marco nodded appreciatively. "We never would have found this without systematic data analysis."

The 5-Why Method with Evidence

Now that the team had gathered evidence, Pedro led them through the systematic 5-Why analysis. This time they were not assumptions, but proven facts.
"Now that we have evidence, we're going to apply the 5-Why method. But based on facts, not assumptions," Pedro announced, as he walked to the whiteboard.

Why 1: The Primary Problem Pedro wrote on the board: "Why do paper jams occur in section C?"

Sandra replied confidently: "The paper roll keeps getting stuck on guide roller 7." She pointed to her observation log. "Evidence: direct observation of 23 jams in three days."

Why 2: The Mechanism "Why does it keep getting stuck?" Pedro continued. Marco Rivera had his measurement results ready. "Too much friction due to static electricity.

Proof: 2.3 kV measured under problem conditions."

Why 3: The Technical Cause "Why does static electricity occur?"

Tom pointed at the electrical schematics. "Guide roller 7 has not been grounded since the overhaul two years ago.

Proof: resistance measurement shows no connection to ground."

Why 4: The Procedural Cause "Why wasn't the grounding restored?"

Jake looked guilty. "No checklist for electrical connections after mechanical maintenance.

"Proof: procedures reviewed – electrical checks missing."

Why 5: The systemic cause "Why don't we have integrated checklists?"

Linda had investigated this as well. "Mechanical and electrical work orders run through different systems.

Proof: EAM system analysis shows no cross-references."

Sandra leaned back in her chair. "Wow. We started with 'machine jams' and end up at our maintenance procedures."
Marco nodded in agreement. "And every step is proven, not guessed."

7.2. Verification Methods in Practice

The InduPack team had applied several verification methods, each with its own strength. It is important to understand when to use which method.

Direct Observation

The most powerful verification method is often also the simplest: observing the problem as it occurs. Sandra's systematic observations yielded crucial insights that no data analysis could have provided.
The observation form she used contained:

- Time of day
- Paper type before/after changeover
- Weather conditions outside
- Number of people on shift
- Preceding sounds/vibrations
- Exact location of jam
- Time to Resolution

By systematically recording this data, it transformed Sandra's experience from 'feeling' to objective data.

Controlled experiments

Marco's experimental approach demonstrated the power of systematic testing. By changing one variable at a time, while keeping all others constant, he was able to establish cause-and-effect relationships with a high degree of certainty.
His test protocol was simple but effective:

- Measure reference situation without changes
- Change one parameter (for example, humidity)
- Observe effect during fixed period
- Document results with timestamp
- Repeat test under different conditions
- Compare results and draw conclusions

"The trick is to be patient," Marco explained. "You want answers quickly, but reliable experiments take time."

Historical Data Analysis

Linda's statistical approach complemented observation and experimentation well. By searching for patterns in historical data, she could identify correlations that would otherwise remain invisible.
Her spreadsheet analyses showed:

- Correlation between jam frequency and season
- Relationship between wind direction and how often the problem occurs
- Time patterns indicating system-related causes
- Cost distribution across different root cause categories

"Spreadsheets are an underrated tool for RCA," Linda remarked. "You don't need expensive software for powerful analyses."

7.3. From Correlation to Causality

A crucial point in every RCA is the distinction between correlation and causality. Linda taught the team to understand this important difference.

"I found another interesting correlation," Linda reported during one of the final verification sessions. "Jams correlate with outside temperature — lower temperature, more jams."

Marco looked interested. "So cold days cause jams?"
"No!" Linda shook her head vigorously. "That's correlation, not causality. Both are influenced by seasons."
Pedro asked her to explain. "Cold air contains less moisture. When that air is warmed inside the building, its relative humidity drops sharply. It becomes bone dry. The temperature doesn't cause the jams — it's part of a chain: cold weather → lower humidity → more static electricity → jams."
"So we're not looking for direct causes, but for mechanisms?" Tom said, beginning to understand.
Sandra nodded. "That explains why our quick fixes never worked. We were addressing symptoms, not the actual chain of events."

Levels of Evidence: From Weak to Strong

The team learned that not all evidence is equally strong. Pedro explained the hierarchy of evidence strength.
"We've now gathered various types of evidence. But not all evidence is equally strong," Pedro said during the closing verification session.
Linda was curious. "What do you mean?"

"There are four levels of evidence.

Sandra's observations are anecdotal - valuable, but not scientific.

"Your correlations, Linda, are statistical — stronger, but still not causality."

Marco looked at his measurement results. "And my electrical measurements?"
"Experimental evidence - much stronger. You've created controlled conditions and demonstrated cause and effect." Tom was curious about the last level. "And the strongest level?"
"Mechanistic evidence — when you understand the exact mechanism. Tom, your explanation about static electricity causing stick-slip motion, which escalates into jams — that is mechanistic understanding."

Jake saw the picture become complete. "So we have evidence on all four levels for our main conclusion?"
"Exactly! That's why we can be certain that we've found the root causes."

Level 1: Anecdotal evidence Sandra's fifteen years of experience provided valuable clues, but was subjective and difficult to

verify.

Level 2: Correlational evidence Linda's statistical analyses

showed strong relationships (correlation > 0.75) but did not yet prove cause and effect.

Level 3: Experimental evidence Marco's controlled tests demonstrated that changes in grounding had a direct effect on jam frequency.

Level 4: Mechanistic evidence Tom's electrical analyses revealed the exact mechanism: how static electricity leads to stick-slip motion that escalates into complete jams.

The Eureka Moment

In week four, everything fell into place. The team had their breakthrough during what would later become known as 'the Friday afternoon meeting'. Sandra suddenly stood up from her chair. "Wait a minute. If I … let's summarize what we've discovered …"

She walked to the whiteboard and began drawing:

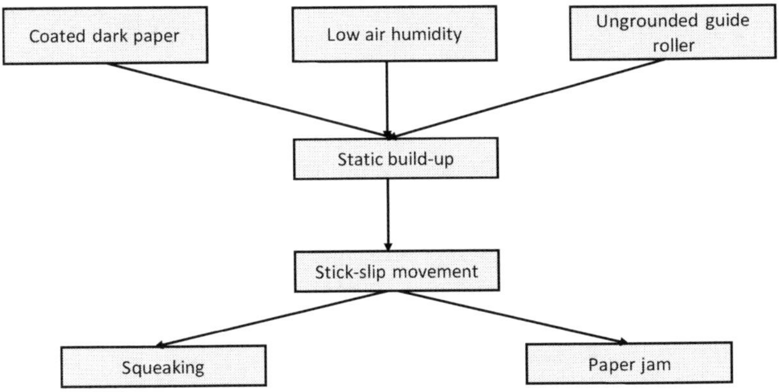

Figure 7-1 Summary of mechanistic evidence by Sandra

Marco jumped up. "And that explains why it only happens early in the morning! At the end of the night, it's coldest outside!"
Tom called out enthusiastically: "And why only at light-to-dark changes!"

Linda pointed to her data. "And the seasonal patterns and the wind direction correlation!"
Jake leaned back. "It all adds up. For the first time, we truly understand what's happening."
Pedro smiled broadly. "This is what systematic RCA can achieve. We've gone from symptoms to the complete mechanism."
Sandra laughed. "I'm almost going to miss the breakdowns. Now I know exactly when they're coming and why."
Marco nodded enthusiastically. "And the great thing is — now that we understand the mechanism, we can come up with targeted solutions."

7.4. Team Evolution Through the Verification Process

It was fascinating to see how the team transformed throughout the verification process. From skeptics they became ambassadors, from individual experts they became an analysis team.

Week 1 — Uncertainty and Resistance

Sandra had doubted it at first: "We've analyzed this problem a hundred times. Nothing works long-term."

Marco was practically minded: "Theory is fine, but I have to solve those jams every day."

Jake stood under pressure: "Management wants quick results, not months of analysis."

Week 2 - First Breakthroughs

Sandra grew enthusiastic when her observations were taken seriously: "For the first time, someone is really listening to what I see every day. And we're doing something systematic with it." Marco discovered new patterns: "I have twenty years of experience, but this methodology is showing me things I never would have noticed. That correlation between paper type and time of day - how did we miss that?"

Week 3 - Growing Confidence

Jake saw the value of the systematic approach: "This is the most systematic approach we've ever had. Now I can go to the director with confidence: these are the causes, this is what we're going to do about it."

Tom appreciated the multidisciplinary collaboration: "By working together with Sandra, I learned how the machine behaves in practice. You can't learn that from diagrams."

Week 4 - Full Conviction
Linda realized the value of teamwork: "My statistical models are only as good as my understanding of the process. Without Sandra's observations, I would have been looking for the wrong correlations." The entire team had gained confidence in the methodology and in each other.

Spreadsheets as a Verification Tool

Linda showed the team how a spreadsheet can be a powerful tool for RCA verification, without expensive specialized software.

Decision-Making on Causes to Address

After verification, the InduPack team had identified three proven root causes. Now they had to decide which ones to address and in what order.

The three proven root causes:

1. Missing ground on guide roller 7 (mechanical)
2. Low air humidity 6.00-8.00 (environment)
3. Insufficient procedures for product changeover (method)

Documentation of the Verification Process

For each hypothesis, the team systematically documented the evidence. This documentation later became essential for knowledge transfer and management reporting.

Documentation format per hypothesis:

Hypothesis: Missing grounding causes static electricity
Expected if true: Electrical measurements show voltage on guide roller
Expected if false: No electrical activity measurable
Test performed: Multimeter measurements during various production conditions
Result: 2.3 kV voltage measured, discharges every 45 seconds
Evidence level: Experimental (level 3)
Conclusion: PROVEN - root cause identified

This systematic documentation had multiple advantages. It forced the team to clearly articulate what they expected to find. It made review and validation by others possible. And it created an audit trail for management and future reference.

Timestamp Documents

All observations received timestamps, photos where possible, and were cross-referenced with production logs. Sandra's observation forms were digitally recorded with handwritten notes scanned and stored.

Expert Validation External experts assessed the findings. An electrical engineer confirmed Tom Davis's analysis of the static electricity. A process specialist validated Linda's statistical correlations.
This documentation was later used for knowledge transfer to other product lines, management reporting on ROI, training of new team members, and audit compliance.

7.5. Lessons for Other RCA Projects

The InduPack project yielded valuable lessons for future RCA initiatives.
Team composition crucial. The combination of practical experience (Sandra), technical expertise (Marco, Tom), analytical skills (Linda), and management support (Jake) under the facilitation of an experienced reliability engineer (Pedro) proved ideal.

Systematic verification pays off. The distinction between guessing and knowing made the difference between a solution that works and one that fails. Each of the fifteen original hypotheses was systematically tested.

Combining Different Levels of Evidence. Anecdotal evidence provided direction, correlational evidence revealed patterns, experimental evidence proved causality, and mechanistic evidence gave complete understanding. All levels were needed for complete certainty.

Spreadsheets as a Democratic Tool. Advanced software is not necessary for effective RCA. Spreadsheets offer all the functionality that most teams need for verification and analysis.

Patient, systematic approach. The temptation to jump to conclusions quickly is great, especially under management pressure. But systematic verification prevents costly mistakes and delivers lasting solutions.

From Verification to Implementation

Now that you have systematically verified your hypotheses and identified proven causes, you can move on to developing structural solutions. The verification process has brought you from fifteen suspects to three proven root causes. In Chapter 8, you will learn how to develop solutions for these root causes that not only eliminate the symptom, but definitively remove the cause.

8. Developing Solutions

8.1. From Proven Causes to Structural Solutions

Now that you have verified the root causes of your failure problem, you can begin working on developing solutions that not only eliminate the symptom but definitively remove the cause. This is the moment when weeks of analysis are converted into concrete action.
Pedro stood at InduPack Solutions at this crucial phase. His RCA team had identified three proven root causes for the conveyor belt jams: missing grounding of guide roller 7, low humidity during the morning shift, and insufficient procedures during product changeovers. Now they had to move from this knowledge to working solutions.

"We've proven what the causes are," Pedro said, opening the brainstorming session. "Now we're going to develop solutions that will make these 78 jams per month a thing of the past. Sandra, you know the day-to-day reality. Marco and Tom, you know what's technically possible. Linda, you keep an eye on the numbers. Jake, you keep track of the budget and time constraints. Together we're going to turn a $240,000 loss into structural profit."

Systematic Brainstorming for Solution Directions

Effective brainstorming for RCA solutions requires input from various disciplines. Operators see daily pain points, technicians know mechanical possibilities, engineers understand system interactions, and managers oversee budget and priorities.
At InduPack, the team used the round-robin technique: each team member took turns sharing ideas without criticism.

Quality Comes from Quantity

"Okay team, we're going to brainstorm, but in a systematic way," Pedro began. "Rule one: no criticism while gathering ideas. Rule two: at least five concrete solutions per root cause."

Sandra frowned. "Five? For that grounding I can only think of one—just connect the wire."
Marco shook his head. "Hold on. We could also use antistatic rollers to replace rollers 5-9. Or replace the entire section C with static-free components."
Tom nodded enthusiastically. "Or install an active discharge system with ionization units. Or adjust the paper properties by using a different supplier."
Linda added: "Or control the environmental conditions with 3-4 industrial humidifiers, so that static buildup can no longer happen."
Jake smiled. "See? From one to five solutions in two minutes."
For each root cause, the team systematically collected concrete options:

Grounding Solutions:

- Direct grounding via copper wire to main ground ($ 1,200 material + 8 hours installation)
- Install antistatic rollers to replace rollers 5–9 ($4,000 + 2 days downtime)
- Install active ionization discharge system ($3,500 + daily monitoring)
- Section C completely replaced with static-free components ($12,000 + one week of downtime)
- Environmental control via humidity between 45-55% ($6,000 humidifiers + monthly calibration)

Humidity Solutions:

- HVAC system adjustment for constant 45-55% RH in paper line room ($8,000 + three days installation)
- Local humidification via 2-3 industrial units near critical zones ($4,500 + daily refilling)
- Adjust production timing to periods with naturally higher humidity levels (no cost + scheduling adjustment)
- Seasonal planning with critical products only during stable months (no installation costs, but higher inventory costs)

- Replace paper grades with less moisture-sensitive alternatives (supplier negotiation + quality testing)

Procedural Solutions:

- Comprehensive training for all shifts in standardized changeover sequence ($2,000 + 40 hours of training)
- Laminated checklists at each machine with 8-step procedure ($300 + monthly updates)
- Visual aids via photos of correct settings per paper type ($550 + quarterly updates)
- Cross-training, so that every shift has at least two experienced changers ($3,000 + 80 hours of training)
- Supervision intensification with dedicated shift escorts during critical moments ($15,000/year in additional labor costs)

"Fifteen concrete solution directions in one hour," Pedro said with satisfaction. "Now we're going to evaluate them, but first, even more creative thinking."

8.2. SCAMPER Methodology for RCA Solutions

Linda suggested thinking even more creatively through systematic transformation questions. "Can we think even more creatively? I know a methodology: SCAMPER."
Jake looked questioningly. "SCAMPER?"
"Substitute, Combine, Adapt, Modify, Put to Other Use, Eliminate, Reorganize," Linda said. "Each letter represents a different way of thinking."
The SCAMPER technique stimulates systematic thinking through transformation questions:

Substitute: What can we replace?

Marco said, "Instead of metal guide rollers, use ceramic rollers that don't build up static charge."
Tom: "Replace the DC motors with AC motors with variable frequency drives for smoother movement."

Combining: What Can We Bring Together?

Sandra: "Combining product changeover procedure with manual belt tension adjustment and antistatic spray"
Linda: "Linking humidity measurements every two hours to automatic ventilation adjustments"

Adapt: What can we modify?

Jake: "Production changeovers only during shifts with humidity above 40%"
Marco: "Manually adjust belt tension based on paper type according to the new reference table"

Modify: What can we dimension differently?

Tom: "Smaller, hourly humidity measurements for better control instead of daily"
Sandra: "Longer intervals of six hours between product changeovers to stabilize the system"

Put to other use: What else can we use this for?

Jake: "The night shift can do humidity measurements - they're less busy anyway"
Sandra: "Use the product changeover moment for a quick belt tension check and preventive inspection"

Eliminate: What can we leave out?

Tom asked: "Can we eliminate guide roller 7 entirely and route the paper via an alternative path?"
Marco: "Eliminate manual product changeovers by keeping 2x as much inventory of the same paper type"

Reorganizing: What Can We Arrange Differently?

Linda: "Reverse the product sequence — start the day with the least critical paper grades when humidity is low"
Jake: "Shift maintenance timing to periods of optimal humidity between 10:00–14:00"

"Now we have 31 concrete solution directions," Pedro concluded. "From conservative to revolutionary. Time to evaluate systematically."

8.3. Evaluation Criteria and Decision Matrices

Developing a Weighted Evaluation Matrix

Jake led the systematic evaluation of all options. "We can't implement all 31 solutions. We need to choose based on clear criteria."
Pedro nodded. "We use a weighted evaluation matrix. We define criteria and assign each criterion a weight based on what truly matters for InduPack."
The team had an intense discussion about priorities:
Linda: "Effectiveness should carry a lot of weight — does it really solve the problem?"
Tom said, "Safety is essential — with static electricity, the solution simply has to work."
Marco: "Costs are important, but not leading. A cheap solution that half works ends up costing more." Sandra: "Implementation speed matters - management wants to see results."
Jake said, "And integration with existing systems — we don't want chaos."
After this discussion, the team established the weights and systematically evaluated each solution:

Criterion and Weight:

- Effectiveness: 50% (primary goal: eliminate jams)
- Implementation costs: 20% (budget impact management)
- Operational costs: 20% (long-term sustainability)
- Implementation speed: 10% (urgency of production loss)
- Integration ease: 5% (existing systems)

Assessing Effectiveness

For each solution, the team estimated the expected reduction in jam frequency and translated this into scores:

- Restore grounding: 70-80% reduction (score 8/10)

- Increasing humidity: 60-75% reduction (score 7/10)
- Improving procedures: 40-60% reduction (score 6/10)
- Combination of all three: 85-95% reduction (score 9/10)

Cost Realism

Linda made realistic cost estimates including:

- Direct costs (material, labor, subcontracting)
- Indirect costs (production loss during implementation)
- Opportunity costs (alternative projects that are deferred)
- Unforeseen costs (15% buffer for complex solutions)

Spreadsheet-Based Scoring Matrix

The team used a spreadsheet to systematically score all criteria on a scale of 1–10, with each score multiplied by the criterion weight. The weighted scores were added together to produce a total score per solution.

Solution	Implementation costs (×20%)	Operating costs (×20%)	Effectiveness (×50%)	Implementation speed (×10%)	Ease of integration (×5%)	WEIGHTED TOTAL
Direct grounding (copper wire)	9 (1,80)	9 (1,80)	8 (4,00)	9 (0,90)	8 (0,40)	**8,9**
Modify HVAC system	6 (1,20)	8 (1,60)	7 (3,50)	6 (0,60)	7 (0,35)	**7,25**
Training all shifts	8 (1,60)	7 (1,40)	6 (3,00)	8 (0,80)	9 (0,45)	**7,25**
Antistatic rollers	7 (1,40)	9 (1,80)	7 (3,50)	5 (0,50)	6 (0,30)	**7,5**
Local humidification (units)	7 (1,40)	6 (1,20)	7 (3,50)	7 (0,70)	8 (0,40)	**7,2**
Laminated checklists	10 (2,00)	8 (1,60)	5 (2,50)	10 (1,00)	10 (0,50)	**7,6**
Combination: grounding + HVAC + training	4 (0,80)	7 (1,40)	9 (4,50)	3 (0,30)	6 (0,30)	**7,3**

Figure 8-1 Weighting matrix for solutions

Jake looked at the results. "The grounding solution scores highest with 8.90 points. That's consistent with our analysis: it's the most effective, affordable, quick to implement, and technically straightforward."

Linda nodded in agreement. "And the combination of all three solutions scores lower precisely because of the high costs and long implementation time, despite the highest effectiveness."

"Exactly," Marco said. "That's why a matrix like this is valuable. It makes trade-offs transparent. We can now see that an 85% jam reduction with the combination doesn't outweigh the nearly three times higher investment and long implementation time."
Sandra added: "And if the grounding solution turns out to be insufficient, we can always add the HVAC modification later. Then we'll have data to support that decision."

8.4. Cost-Benefit Analysis with ROI Calculation

Systematic Cost Inventory

One-Time Investments

For the grounding solution, Marco inventoried:
Material costs: $1,200 (copper wire, terminals, conduit)
Labor costs: $800 (8 hours of Tom's time at $110/hour)
Downtime costs: $600 (3 hours of production loss at $200/hour)
Total: $2,600

Recurring Operational Costs

Linda calculated annual costs:
Grounding inspection: $300/year (annual electrical check)
Documentation updates: $150/year
Total annual: $450

Benefits Quantification

Direct Savings from Jam Reduction

Historical data showed: 78 jams/month costing a combined $240,000 per year. 70% reduction through grounding =
$168.000 annual savings

Indirect Savings

The team identified additional benefits:
Reduced operator stress: $8,000/year (reduced absenteeism)
Lower maintenance costs: $4,200/year (less wear from emergency stops)

Quality improvement: $6,300/year (fewer defective products)
Total indirect: $18,500/year

Marco whistled admiringly. "And that's just for one solution. If we implement all three?"
Linda kept typing in her spreadsheet. "Combined impact of 85-90% jam reduction. More than $200,000 in annual savings."
Sandra shook her head. "Now I understand why we worked systematically. Management can't ignore these numbers."

8.5. Virtual Testing and Test Phase Methodologies

Test Phase Planning

Marco wanted to test the solutions before full implementation would take place. "Before I spend $2,600, I want proof that it works. Can we do a test phase?"
Pedro nodded enthusiastically. "Absolutely! We test every solution before we commit."
Tom said, "I can temporarily install a test ground. If it works, we'll make it permanent."
Marco said, "And I want accelerated testing — 72 hours of continuous monitoring under various conditions."
Sandra: "Can we also do A/B testing? One shift with grounding, one shift without?"
Jake asked, "How much time does testing take?"
Tom: "For grounding: one week setup, two weeks testing, one week evaluation."

Accelerated Testing Protocols

Before implementation, Marco tested the grounding solution under extreme conditions. This way he could map out in a short time the problems that would otherwise occur over a longer period:

- 72-hour continuous test at low humidity (25%)
- Accelerated product changeovers (every 30 minutes vs. the normal 4 hours)

- Maximum belt speed during coated dark paper

"The accelerated test simulates six months of normal operation in three days," Marco explained. "Result: zero jams with grounding vs twelve jams without grounding."

A/B Testing for Procedural Solutions

Sandra tested two different product changeover procedures in parallel:

1. Procedure A: *Traditional Replacement Without Additional Steps*
2. Procedure B: New procedure with antistatic spray and 5-minute stabilization

Over four weeks: Procedure A had 23 jams during product changeovers, Procedure B had three jams. A clearly measurable improvement.

8.6. Change Management

Jake realized that technical solutions alone are not enough. "Hold on. We have technical solutions, but how do we make sure people actually use them? Those new procedures, for example."
Pedro smiled. "Good point! Technical solutions are easy. Changing people is hard. That's why we use the Four A's."
Sandra looked up questioningly. "Four A's?"
"Accept, Align, Agree, Address," Pedro said. "Every A-aspect has to be right for successful implementation."

Accepting: Creating Awareness and Urgency

"People don't like to change," Pedro explained. "They're used to their way of working. It's not perfect, but it keeps the plant running. New ways always come with risks, plus they have to get used to them themselves. That's why people first need to truly feel that change is necessary."
Jake made the costs concretely visible: "These 78 jams per month cost us more than a new BMW every month. Look, here's a photo of a 5 Series — that's what we're losing because of this problem."
Sandra told her story: "I run back and forth every day for those jams. My step counter shows that I take 3,000 extra steps per day just because of

this problem. After 15 years here, I'm getting tired of the same frustration over and over."
"Important," Pedro emphasized, "these improvements cost no one their job. We use the freed-up time to make the plant even more profitable and to strengthen our competitive position."

Alignment: Participatory Solution Design

"Alignment ensures that everyone feels heard in how the change is going to take place," Pedro explained. "You can't have every operator and technician contribute their thinking, but you do need representatives from each group. The best choices are people who are seen by their colleagues as 'people who know what they're talking about.'"

Pedro organized working groups where representatives from each discipline could provide input on advantages, disadvantages, and elements that absolutely could not be missing from the new approach.
Sandra, speaking as an experienced operator, spoke on behalf of the production staff: "That new procedure needs to stay practical. No complicated steps during a busy product changeover."
"And those grounding checks need to fit into our existing rounds," Marco said. "Not another separate task on top of everything else."

Agreeing: Concrete Implementation Agreements

"Making agreements means properly managing processes, people, and resources," Jake explained. "Procedures need to be clear, people trained, time scheduled, materials available, work orders updated. If people need to clean something, for example, it needs to be clear what supplies they use, where those are stored, and who reorders them when they run out."
For the grounding solution, Tom arranged: ordering materials, scheduling installation during a maintenance shutdown, electrical measurements, and training for operators in visual inspections.
Sandra and Jake worked out the new procedures: a pilot on one shift, training for all employees, and integration into existing work orders, so it wouldn't create any extra administrative burden.

Addressing: Feedback and Recognition

"Improvements often lose their impact because management thinks: 'We've implemented this now, we don't need to look at it anymore,'" Pedro warned. "Without constant attention, you easily slip back into old habits."

Pedro introduced weekly check-ins: "How did this week go? What went well? What are we running into?" He deliberately gave more compliments for good performance than criticism for problems.

Linda set up a dashboard that tracked jam frequency per shift, but used it primarily to highlight positive trends. "Shift B has had three jam-free days in a row! That's the best performance of the quarter."

8.7. Implementation Strategy with a Phased Approach

Pedro didn't want to implement all the solutions at once. "A phased approach gives us quick wins and builds trust," he explained to Jake.

They started with the grounding solution, because it was the fastest to implement and had the greatest impact. Tom installed the grounding over a weekend, and within a week they already saw the jam frequency cut in half.

"Perfect," Pedro said, smiling as he saw the first results. "Now the team has confidence for phase two: the new procedures." Sandra trained only the morning shift first, and only after two weeks of successful results did they expand to all shifts.

The humidity solution came last, because it required the largest investment. But thanks to the earlier successes, management was willing to invest $8,000 in HVAC modifications.

Results Measurement and Verification

After three months of implementation, Linda was able to present the results. "From 78 jams per month down to 17," she reported proudly. "That's a 78% reduction, and the average production loss per jam has also dropped from 45 to 23 minutes."

The team had not only achieved the technical goals, but had also improved the atmosphere. Sandra beamed: "Finally I can just do my job

without constantly running around for jams. And the new procedures make product changeovers much calmer."
Marco nodded in agreement: "The systematic approach has proven that it works. Now we can use the same method for other problems on other lines."
The investment of $5,000 yielded an annual savings of $200,000.

Wrap-Up and Looking Ahead

Pedro summarized as the team evaluated their implementation: "We've proven that systematic RCA doesn't just solve problems, but also teaches the organization to address problems systematically."
Sandra rubbed her hands together. "And now we can use this approach for the next problem on line 5."
Marco picked up his tools. "With this methodology, we'll get those conveyor belt issues sorted out there as well."
Jake looked at the larger organization. "From one problem to organization-wide improvement."
Pedro wrapped up: "That is the power of systematic Root Cause Analysis. Next week we start on line 5."

Epilogue: Putting Root Cause Analysis in Maintenance in Perspective

Pedro, eighteen months later

Pedro walked through the production floor of InduPack Solutions, a place he had come to know eighteen months earlier as a collection of unsolved problems and frustration. Now the dashboard showed numbers that made all the difference: 73% fewer failures, 68% lower failure costs, average repair time down from 45 to 18 minutes.

"Nine times out of ten when I come into the plant, it's for proactive improvement rather than a crisis," he said.

Conveyor belt line three – once responsible for 78 jams per month and $240,000 in annual losses – now runs with just 12 incidents per month. The systematic RCA approach had worked: grounding restored, humidity controlled, procedures standardized.

But the eighteen months had also brought new insights. Some failures were symptoms of deeper problems: wrong design choices at the time of purchase, inadequate maintenance plans, insufficient training of operators.

"Systematic failure reduction solves a lot," Pedro acknowledged. "But I also saw where it stops."

What Structurally Reducing Failures Does and Does Not Solve

Pedro's transformation demonstrates the power of systematic Root Cause Analysis. You now know how to define problems sharply, systematically gather information, apply the 5M method and Fault Tree, and move from suspicions to proven solutions.

But as Pedro discovered, challenges remain:

Preventive maintenance: They regularly analyze problems and discover that these arise because they are performing the wrong preventive maintenance. Setting up the maintenance correctly for the entire company all at once is better.

Work Organization: A perfect solution is of no help if implementation gets bogged down in chaotic workflows. Planning and executing thousands of work orders requires a systematic approach that goes beyond knowing 'what' the solution is.

Design flaws: Some problems are already embedded in the DNA of the asset. Retrofitting after the fact is expensive and complicated – it would have been better to build reliability in during the design phase.

Assembly and execution: Even the best technical solution fails if the execution is flawed. Incorrect alignment, improper torque values, or contamination during assembly will undermine your RCA efforts.

Spare parts: You know exactly which part needs to be replaced, but if it has a four-week lead time while the machine needs to go down tomorrow...

Operator involvement: Many small problems can be prevented by operators themselves or detected at an early stage – if they are trained and motivated to do so.

Different Routes to the Same Goal

Pedro began, like many organizations, with structural failure reduction. It quickly delivered measurable results and created support for further improvements. But it is not the only starting point.

Some organizations start with preventive maintenance – proactively preventing failures from occurring in the first place. This works especially well in companies with many critical assets where downtime is unacceptable. It creates stability and predictability, which makes later improvements easier.

Others start with workflow control because their maintenance department is drowning in chaos and ad-hoc work. Or with autonomous maintenance because motivated operators can make the difference with simple machines.

The Nine Disciplines for Structurally Reliable Assets

The challenges that remain — from work organization to preventive maintenance, from inventory management to design flaws — each call for their own systematic approach. This book series covers nine disciplines that together answer the question: "What must an organization do to make its technical assets perform as well as possible?"

1. Maintenance Management – The Overarching Framework
2. Planning and Scheduling in Maintenance– Efficient Organization of Work Activities
3. Troubleshooting Breakdowns– Effective Intervention in Acute Problems
4. Root Cause Analysis in Maintenance– Addressing Recurring Problems
5. Preventive Maintenance – Proactive Asset Care
6. Spare Parts Management – Optimal Inventory Control
7. Precision Maintenance– Flawless Installation and Overhaul
8. Engineering for Reliability – Reliability During Design
9. Autonomous Maintenance – Operator Involvement

The nine books follow a logical structure in which each book builds on knowledge from earlier parts. The implementation order in your organization may differ, however — start where the pain is greatest and where you can achieve results quickly. In the introduction of this book you will find a complete overview of all parts and how they relate to one another.

More information about the complete series
For a current overview of all books, additional information and ordering options, visit:

vadeo.nl/books.html

Every discipline you add makes the others more effective. They reinforce each other. Good work flow control ensures you have time for precision work. Preventive maintenance works better when operators detect deviations early. Reliability engineering prevents problems that you would otherwise have to address structurally. The order in which you implement them depends on what your organization needs most.

Your Next Step

You have now learned a methodology that works. InduPack's results – and the thousands of organizations that apply similar approaches – prove that systematic failure reduction transforms organizations. The question is: do you stop here, or do you continue building?

For now: start with what you have learned in this book. Set up an effective registration system. Start daily failure reviews. Use Pareto analysis to set priorities. Apply the 5M method and Fault Tree. Verify your hypotheses systematically. Measure your results. Celebrate your successes.

Pedro's advice: "Start where the pain is. Measure results. And if you discover that other aspects are holding you back, there are also systematic solutions for that. We started with failure reduction because we were drowning in acute problems. That worked for us. But I know companies that started with preventive maintenance and were just as successful. It's not about the sequence, but about the methodology."

Appendix A: Template for failure report

FAILURE REPORT		Date	Department	Asset name	ID No.
Title					
Technician(s)			Operators		
Failure start	Failure end	Production downtime	Repair start	Repair end	Hours worked
Impact	Safety [ ]	Environment [ ]	Quality [ ]	Regulatory [ ]	
Symptoms					
Direct cause					
Repair performed					
Settings / parameters after correction					
5 x Why 1			Materials description	Quantity	No.
2					
3					
4					
5					
Follow-up actions required?					
Remarks / lessons learned					

Figures

Glossary — Terminology and Acronyms

5M Model (Ishikawa) Systematic categorization of possible causes into five main groups: People (competence, procedures, communication), Machine (mechanical, electrical, calibration), Environment (temperature, humidity, contamination), Material (quality, properties, supplier), Method (procedures, standards, processes).

5W+1H Checklist Systematic questions for sharp problem definition: What (exact problem), Where (location), When (timing/circumstances), Who (affected/involved), Why (impact/importance), How (recognizable).

5-Why Method A technique that asks 'why?' five times to move from symptom to root cause; effective for problems with a linear causal chain.

Action Codes Standardized codes that describe what was done: AC01-Replaced, AC02-Repaired, AC03-Cleaned, AC04-Adjusted, AC05-Lubricated, AC06-Tightened, AC07-Reset, AC08-Temporarily repaired.

Anecdotal Evidence (Level 1) The weakest form of evidence, based on observations and experiences; valuable as a starting point but insufficient for costly solutions.

Barrier Analysis A methodology that asks which lines of defense should have prevented the cause from leading to the problem, and why those barriers failed.

Cause Codes Standardized codes that describe why a failure occurred: OC01-Normal wear, OC02-Operating error, OC03-Maintenance deficiency, OC04-External cause, OC05-Material defect, OC06-Assembly error, OC07-Overload, OC08-Unknown.

Change Analysis A research method that focuses on what changed before a problem began; often the fastest route to identifying the trigger of new problems.

Chronic Failure High frequency, low impact per incident; often accepted as 'normal' but cumulative impact can be greater than sporadic failures (example: 5 minutes of loss every hour = 300 minutes per month).

Confirmation Bias A psychological pitfall in which information that confirms initial ideas is primarily sought; prevented through systematic methodology and a diverse team.

Correlational Evidence (Level 2) A statistical relationship between variables; proves association but not yet a cause-and-effect relationship.

Cost-Benefit Analysis A systematic comparison of total costs (one-time + recurring) versus total benefits (direct + indirect) with ROI calculation; supports investment decisions.

Daily Failure Review Meeting A brief daily meeting (max 20 minutes) for initial triage between acute and structural problems; gathers information, identifies patterns, and delivers quick wins.

Damage Codes Standardized codes that describe what is visible: SC01-Failed, SC02-Leak, SC03-Loose, SC04-Broken off, SC05-Blocked/Contaminated, SC06-Abnormal wear, SC07-Makes noise, SC08-Vibrates/Moves, SC09-Overheated, SC10-Performs poorly.

Direct Costs Visible repair costs: parts, labor hours, outsourcing; typically only 15–25% of total failure costs.

EAM (Enterprise Asset Management) Software for planning, execution, and registration of maintenance and spare parts management.

Early Signaling Operators report minor deviations before they escalate into failures; prevention through early intervention.

Event Tree Analysis (ETA) A forward-looking analysis that starts at an initiating event and maps all possible consequences; used for safety and risk analyses.

Experimental Evidence (Level 3) A controlled test in which one variable is changed and a consistent effect is observed; a strong basis for solutions.

Failure (broad definition) Any situation in which an asset fails to meet expectations: complete failure, reduced performance, quality problems, safety risks, or deviations from normal operation.

Failure Report Structured documentation of a failure with at minimum: asset, date/time, reporter/technician, duration/impact, safety/environment, symptoms, cause, root causes, solution, follow-up actions.

Fault Tree Analysis (FTA) A backward-looking analysis that systematically moves from symptom to deeper causes via a logical tree structure; it visualizes causal chains and parallel causes across multiple levels.

Hidden Costs Indirect costs of failures that are often 4–15 times higher than direct repair costs: production loss, rejected products, overtime, missed deliveries, customer compensation, stress and absenteeism.

Impact-Investment Matrix A decision framework with two axes (impact/investment) that classifies solutions into four quadrants: address immediately, regular work order, systematic RCA, consciously defer.

Improvement Team A multidisciplinary team (typically 5–8 people) for an RCA project: facilitator, operators, maintenance, technical experts, management; combines hands-on experience with analytical skills.

Is/Is-Not Analysis An analysis method that systematically defines where a problem DOES and DOES NOT occur; helps eliminate causes and narrow the search area by identifying contrasts.

Jumping to Conclusions Assuming the cause too quickly without thoroughly verifying it; leads to solutions that do not work or only have a temporary effect.

Lack of Focus A common mistake where too many problems are addressed simultaneously; results in fragmentation and few concrete results.

Mechanistic Evidence (Level 4) The strongest form of evidence, in which the exact mechanism is understood; complete understanding of the causal chain.

No Dedicated Resources A situation in which the same people are responsible for both acute and structural problems; urgent work always crowds out structural work.

OEE (Overall Equipment Effectiveness) A performance measurement consisting of Availability × Performance × Quality, expressed as a percentage.

Pareto Chart A bar chart in which causes are sorted from highest to lowest impact, with a cumulative line showing what percentage of total losses is caused by the top causes.

Pareto Principle (80/20 Rule) A principle whereby 20% of the problems cause 80% of the impact; fundamental for prioritization by focusing on the 'vital few' rather than spreading energy across many problems.

Phased Implementation Step-by-step rollout starting with quick wins to build confidence, then moving to more complex solutions; prevents chaos and builds commitment.

Primary Effect The main problem you want to solve; can be technical (component fails) or operational (production loss) depending on which yields the most effective solution.

Quick Win A problem where the solution is clear and requires little time or money; addressed directly without extensive analysis for quick results and team confidence.

Reliability Engineer An engineer responsible for the systematic improvement of asset reliability, primarily through structural failure reduction; facilitates RCA projects and ensures a methodical approach.

Resource-Based Prioritization Realistic planning based on available capacity; typically 1–5 structural improvement projects running in parallel, depending on team size and problem complexity.

Risk Matrix A decision tool that combines probability and impact for prioritization; safety/compliance receive higher priority independent of financial impact.

ROI (Return on Investment) Return on investment calculated as a percentage, expressed as annual savings divided by the investment.

Root Cause Analysis (RCA) A systematic methodology for identifying and eliminating the fundamental causes of failures, rather than treating symptoms. The goal is to resolve problems permanently so they do not recur.

SCAMPER Methodology A creative brainstorming technique for solutions: Substitute, Combine, Adapt, Modify, Put to other use, Eliminate, Reorganize.

Silo Mentality An organizational culture in which disciplines (mechanical, electrical, operations) operate as separate islands without cross-functional collaboration; a fundamental barrier to structural improvement.

Six Loss Types (OEE) Categorization of losses: 1) Downtime losses (unplanned/planned), 2) Speed losses (reduced speed, minor stops), 3) Quality losses (startup rejects, production rejects).

Sporadic Failure Low frequency, high impact per incident; rare but significant events with high visibility (example: main bearing fails once per year, 2 days of downtime).

Structural Improvement An approach that eliminates the underlying cause of a problem rather than merely treating symptoms; results in sustainable solutions that prevent recurrence.

Symptom Treatment A superficial approach that resolves immediate consequences without addressing underlying causes; results in recurring problems.

Symptom versus Cause A symptom is an observable problem (machine stops); a cause is why it happens (sensor failure); a root cause is why that cause exists (no preventive maintenance).

The Four A's (Change Management) A systematic approach for sustainable implementation: Acceptance (creating urgency), Alignment (participatory design), Agreement (concrete arrangements), Accountability (feedback and recognition).

Total Failure Costs The sum of direct and hidden costs; a realistic estimate for prioritization and investment decisions.

Triangulation A verification method in which every significant finding is verified through three different independent sources before being deemed proven.

Verification Process Systematic evidence gathering for hypotheses through: direct observation, controlled experiments, historical data analysis, expert validation; transforms assumptions into proven facts.

Virtual Testing A test phase for solutions before full implementation: temporary installation, accelerated tests (72 hours simulates 6 months), A/B testing; proves effectiveness before major investment.

Weighted Evaluation Matrix A systematic assessment of solutions against criteria with different weights. For example: effectiveness (50%), implementation costs (20%), operational costs (20%), implementation speed (10%), ease of integration (5%).

Acronyms and Abbreviations

CMMS Computerized Maintenance Management System
EAM Enterprise Asset Management
ETA Event Tree Analysis
FTA Fault Tree Analysis
KPI Key Performance Indicator
MTBF Mean Time Between Failures
MTTR Mean Time To Repair
OEE Overall Equipment Effectiveness
RCA Root Cause Analysis
ROI Return on Investment
SCAMPER Substitute, Combine, Adapt, Modify, Put to other use, Eliminate, Reorganize

Made in United States
Orlando, FL
04 November 2026

90209057R00085